Lecture Notes in Mathematics

P9-AQX-010

Alexander Mielke

Hamiltonian and Lagrangian Flows on Center Manifolds

with Applications to Elliptic Variational Problems

Springer-Verlag

Berlin Heidelberg New York
London Paris Tokyo
Hong Kong Barcelona
Budapest

Author

Alexander Mielke
Mathematisches Institut A
Universität Stuttgart
Pfaffenwaldring 57
W-7000 Stuttgart 80, FRG

Mathematics Subject Classification (1991): 58F05, 70H30, 35J50, 34C30, 57S20, 73C50, 76B25

ISBN 3-540-54710-X Springer-Verlag Berlin Heidelberg New York
ISBN 0-387-54710-X Springer-Verlag New York Berlin Heidelberg

© Springer-Verlag Berlin Heidelberg 1991
Printed in Germany

Typesetting: Camera ready by author
Printing and binding: Druckhaus Beltz, Hemsbach/Bergstr.
46/3140-543210 - Printed on acid-free paper

Für Bärbel, Annemarie, Lisabet und Frieder

Preface

It is the aim of this work to establish connections between three fields which seem only loosely related from the usual point of view. These fields are described by the following terms: Hamiltonian and Lagrangian systems, center manifold reduction, and elliptic variational problems. All three topics have had a period of fast development within the last two decades; and the interrelations have grown considerably. Here we want to consider just one facet at the intersection of all three fields, namely the implications of center manifold theory to the study of variational problems. The main tool for the analysis is the Hamiltonian point of view.

The original motivation for this work derives from my interest in Saint–Venant's problem. Having in mind the center manifold approach and hearing about the Galerkin or projection method to derive rod models, I thought it worthwhile to study the connections between these two reduction procedures. However, it soon turned out that the tools to be developed involved fairly general ideas in Hamiltonian and Lagrangian system theory. I realized that many of the necessary results are known but spread over many sources or hidden in very abstract clothing. Often, the special needs for our objective were not directly covered. Thus, the plan evolved to write the abstract Part I on Hamiltonian and Lagrangian systems as self–contained as possible.

I made a controversial decision concerning the use of methods and notations from differential geometry. Since many applied researchers are not familiar with differential forms and coordinate free analysis on manifolds, I avoided these tools as much as possible. On the one hand the center manifold is a local object and can be described by one coordinate chart; but on the other hand it is nevertheless a manifold and in order to define a Hamiltonian system on it, the use of differential forms is absolutely necessary. Additionally, the concept of Lie groups involves global manifolds. In this conflict I was guided by the idea to enable nonspecialists (I am one of them) to follow the analysis, give them a first contact to methods in analysis on manifolds and symplectic geometry, and, finally, to motivate them to study these methods in their own right. The reader should judge how far this goal is reached.

To appreciate the abstract methods of Part I it is strongly recommended to get involved in some of the applications studied in Part II (the easiest one is given in Section 8.2). For classical Hamiltonian systems the center manifold does not play an important role, since most systems are oscillatory, which leads to high–dimensional center manifolds. However, center manifolds in elliptic problems in cylinders, first found by Kirchgässner

[Ki82], reduce an infinite–dimensional system to a finite–dimensional one. In nonlinear elasticity one application is the Saint–Venant problem concerning the static deformations of a very long prismatic body. In mechanics this situation is modelled by rod theory which is an ordinary differential equation replacing the equilibrium equations of three-dimensional nonlinear elasticity. Such rod models can now be justified by the center manifold approach; and it was the question of finding the variational structure of these rod equations which brought my attention to this exciting field.

The research reported here was initiated during a one–year stay at Cornell University, where I had numerous, very stimulating discussions with Phil Holmes. I am very grateful to him. Additionally, I would like to thank all the persons who supplied me with interesting hints and helpful comments during the development of this work. Especially, I want to mention M. Beyer, T. Healey, H. Hofer, G. Iooss, J. Marsden, J. Moser, P. Slodowy, and E. Zehnder. Finally, my thanks extend to my former advisor and teacher K. Kirchgässner, who played a major role in my education in mathematics and their applications and who encouraged me whenever needed.

Acknowledgements

The research was supported by the Mathematical Sciences Institute (MSI), Cornell University, and through the grant Ki 131/4-1 of the Deutsche Forschungsgemeinschaft (DFG), Bonn.

Stuttgart, May 1991   *Alexander Mielke*

Contents

Chapter 1

Introduction

In applications, very often complex mathematical models are developed to describe certain phenomena in nature. However, in some circumstances it turns out that only a few critical modes are relevant for the basic effects. This is especially the case when systems with an equilibrium state near the threshold of instability are considered. Then it is desirable to have a method to reduce the complex system to a simpler one which only takes into account the amplitudes of the critical modes. The basic requirement for such a reduction procedure is that the reduced model is a faithful representation of the original problem, at least locally. This means that the solution sets of both systems should be in a one–to–one correspondence for the solutions one is interested in.

For steady problems such a method is provided by the Lyapunov–Schmidt reduction, whereas for time–dependent systems the center manifold reduction is available. There are further reduction methods which may not (yet) be justified in a mathematically rigorous way (in the sense of faithfulness), but are still widespread in the applied sciences due to their simplicity and their important results. Examples are the so–called Galerkin approximations [GH83, pg.417] and amplitude or modulation equations of Ginzburg–Landau type [NW69, DES71]. For the latter, initial results for a mathematical justification are given in [CE90, vH90, IM91, Mi91b].

For Hamiltonian and variational methods it is well–known that Galerkin approximations are best done on the Hamiltonian function and on the functional rather than on the associated differential equation. Then, the reduced problem maintains the Hamiltonian or variational structure; see e.g. [Ho86, BB87] for examples in fluid dynamics and [An72, Ow87] for elasticity theory. We will call these methods simply *projection methods*. It is our aim to recover these methods from a mathematical rigorous basis.

The main motivation for the present work is the study of elliptic variational problems on cylindrical domains, in particular Saint–Venant's problem for the deformations of long prismatic bodies (Ch. 11). It was first shown in [Ki82], that elliptic systems in cylinders can be considered as (ill–posed) evolutionary problems with the axial variable as time, and that these problems are accessible by the center manifold reduction. With this tool one is able to construct all solutions on the infinite cylinder which stay close to a given solution being independent of the axial variable. This method was extensively developed in [Fi84, Mi86a, Mi88b, Mi90, IV91]. Here we are not concerned with the technicalities necessary for proving the existence of center manifolds in this context.

Our interest lies in the question, what additional information about the reduced problem on the center manifold can be gained when the original elliptic problem was obtained as Euler–Lagrange equation of a variational problem. The best we can expect is that the reduced problem can be again understood as a reduced variational problem. For example, this is the case for the Lyapunov–Schmidt reduction, see Section 6.1. But for center manifold reduction, the desired result is, in general, false. Fortunately, in most applications it can be shown that the center manifold reduction of a variational problem is again variational. We state the conjecture that for *strongly elliptic* systems this is always true (Section 6.3).

The main idea to approach variational problems in cylinders is to exploit the distinguished role of the axial variable. Starting with the energy density $f = f(y, u, \nabla_y u, \dot{u})$, where $(y, t) \in \Sigma \times I\!R$ with Σ being the cross–section and $\dot{u} = \partial u / \partial t$, we define a Lagrangian

$$L(u, \dot{u}) = \int_\Sigma f(y, u, \nabla_y u, \dot{u}) \, dy$$

where $(u, \dot{u})$ is now assumed to be an element of some function space over the cross–section Σ. Thus, the system can now be considered as a Lagrangian problem in an infinite–dimensional space. In this sense we use the words 'Lagrangian problem' and 'variational problem' synonymously. Associated to this Lagrangian formulation is a Hamiltonian formulation $H(u, v)$ which is be obtained by the Legendre transform $v = \partial L / \partial \dot{u}$. This relation can always be locally inverted due to the ellipticity of the problem.

Now, all the tools of Hamiltonian systems are at our disposal. For instance, bifurcations of solutions, which are periodic in the axial direction, have to obey the more stringend rules of Hamiltonian bifurcation theory. In [BM90] this idea is applied to the bifurcations from the Stokes family in the theory of surface waves.

Center manifolds are mostly studied in cases where they are stable. This explains why they are not used very often in Hamiltonian theory, since there the symmetry of the spectrum does not allow for a stable center manifold. However, the basis of the present work is the observation that the flow on the center manifold of a Hamiltonian system is again described by a reduced Hamiltonian system. Although this fact seems to be well–known in the realm of Hamiltonian theory, one hardly finds references to this; see e.g. [Kl82, Ch.3.1] for the linear case and [Po80, Ch.2E] for the case involving simple eigenvalues only. The only general treatment, the author is aware of, is given in [Mo77].

After recovering canonical coordinates on the center manifold we have to check whether it is possible to do the inverse Legendre transform; then a reduced variational problem is found. We represent this method in the following diagram.

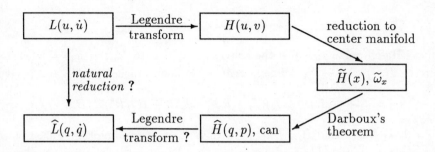

This method constitutes the core of the work presented here. Of course, it can be considered independently of elliptic problems, purely on the level of Lagrangian and Hamiltonian systems. The abstract derivation of the method is done in Part I, and applications to several problems in continuum mechanics are given in Part II.

One remaining open question is to clarify the relations between the full Lagrangian L and the reduced Lagrangian $\widehat{L}$. The most natural relation would be that $\widehat{L}$ equals the restriction of L to the center manifold. If this holds, we use the notion of *natural reduction*. Note that this is exactly the case when the projection (or Galerkin) method yields the same reduced problem as our method. However, the results in this area (Section 6.5) are only preliminary and further research is needed.

On the contents

In Chapter 2 we introduce the necessary notations for Hamiltonian systems on manifolds and then give a basic introduction to the ideas of center manifold theory. Next we treat linear Hamiltonian systems and present some results on linear normal form theory, which are needed later on.

In Chapter 4 we consider center manifolds in Hamiltonian systems. In the first section we give the basic reduction theorem due to Moser [Mo77] and certain related results. In Section 4.2 we generalize the result to Poisson systems as follows. Let X be a Banach space which splits into $X_1 \times X_2$ where X_1 is finite–dimensional. Consider a Hamiltonian system on X given by

$$\begin{pmatrix} \dot{x}_1 \\ \dot{x}_2 \end{pmatrix} = \begin{pmatrix} J_1(x) & J_2(x) \\ J_3(x) & J_4(x) \end{pmatrix} \begin{pmatrix} \mathbf{d}_{x_1} H(x) \\ \mathbf{d}_{x_2} H(x) \end{pmatrix}. \tag{1.1}$$

Here, H is the Hamiltonian and $J = \begin{pmatrix} J_1 & J_2 \\ J_3 & J_4 \end{pmatrix} : X^* \to X$ defines the Poisson bracket $\{F, G\} = \mathbf{d}F(J\mathbf{d}G)$. We assume that (1.1) has a center manifold M_C being tangential to X_1 in the point $x_1 = 0$, i.e. $M_C = \{(x_1, h(x_1)) \in X : x_1 \in U_1 \subset X_1\}$, where the reduction function h satisfies $h(0) = Dh(0) = 0$.

Under the sole assumption that $J_4(0) : X_2^* \to X_2$ has a bounded inverse we show that the flow on M_C is described by

$$\dot{x}_1 = \widetilde{J}(x_1)\mathbf{d}_{x_1}\widetilde{H}(x_1). \tag{1.2}$$

The reduced Hamiltonian $\widetilde{H}$ is just the restriction of H onto M_C: $\widetilde{H}(x_1) = H(x_1, h(x_1))$. The operator $\widetilde{J}(x_1) : X_1^* \to X_1$ defines a reduced Poisson bracket on M_C and is given by

$$\widetilde{J}(x_1) = J_1 + (-J_1B^* + J_2)(J_4 - BJ_2 - J_3B^* + BJ_1B^*)^{-1}(BJ_1 - J_3) \tag{1.3}$$

where $J_i = J_i(x_1, h(x_1))$ and $B = Dh(x_1)$, with B^* being the adjoint. In fact, this reduction method is not restricted to center manifolds but holds for every invariant submanifold with invertible J_4.

In Section 4.3 we treat the method of *flattening of center manifolds* in canonical coordinates. This procedure gives an effective way of calculating the reduced Hamiltonian in canonical coordinates, up to any given order. This method is especially useful for implementation, either numerically or by symbolic manipulations. Finally, we are concerned with the analyticity problem for center manifolds in Hamiltonian systems.

In Chapter 5 systems are studied which are invariant under the action of a Lie group. A short introduction to Lie groups is given in Section 5.1 and the classical *reduction of Hamiltonian system by symmetry* ([Ma81, MR86]) is outlined. Note that the reduction onto a center manifold is completely different from the classical reduction. There a reduced phase space is obtained by factoring with respect to a symmetry action and by restricting onto level surfaces of the corresponding first integrals. The center manifold, however, is characterized only by the dynamical behavior of the solutions lying on it. In particular, in Section 5.2 we show that systems, being invariant under the action of a Lie group, give rise to a reduced system on the center manifold which inherits all the symmetries of the original system. For this purpose we use a slice theorem to decouple the symmetry action. Furthermore we develop special versions of Poincaré's Lemma and Darboux's Theorem for the symmetry case. It follows then that the symmetry reduction can be done after the center manifold reduction.

Variational problems are treated in Chapter 6. First we consider the Lyapunov–Schmidt reduction and the projection method which both have the property that their reduced problem has a natural variational structure. We point out the essential differences to our approach and to get ideas what the desired results for Lagrangian systems are. As a byproduct, we obtain that the center manifold reduction in a gradient system is again a gradient system, but with respect to a non–flat metric.

Then we study the abstract setting of Lagrangian problems and their relation to canonical Hamiltonian systems. The general reduction method proposed above always leads from the original Lagrangian problem, via the Legendre transform and the center manifold reduction, to a reduced Hamiltonian system. Darboux's theorem then supplies canonical coordinates and we would like to transform back into a Lagangian problem. Yet, in general this is not possible, even when all possible canonical coordinates are taken into account. The following condition is necessary and sufficient for the existence of an associated Lagrangian problem. Denoting the linearization of the vector field in (1.2) at $x_1 = 0$ by $K_1 x_1$ this condition is simply given by

$$\dim(\text{kernel } K_1) \leq \tfrac{1}{2}\dim X_1. \tag{1.4}$$

This result is very useful, since it makes the sometimes cumbersome calculation of the quadratic part of $\tilde{H}$ superfluous. Using this condition the reduction procedure can be completed as given in the above diagram. The important question of natural reduction is discussed in Section 6.5. However, by now only a relaxed version of it is well understood.

Moreover, we consider symmetric Lagrangian systems which may have a *relative equilibrium* with respect to the action of a Lie group. Using augmented Hamiltonians we are able to construct a center manifold close to the relative equilibrium, such that the reduced Lagrangian system is again invariant under the reduced action. This theory will be basic for the understanding of Saint–Venant's problem.

In Chapter 7 we are concerned with nonautonomous Hamiltonian or Lagrangian systems. Under suitable uniformity conditions on the time–dependence, the existence of a time–dependent center manifold is known. Then the reduced symplectic structure depends on time also. Using an especially adapted version of Darboux's theorem we find canonical coordinates while keeping the time variable unchanged and preserving the qualitative time–dependence like (quasi–) periodicity. Moreover, for an autonomous system under a nonautonomous perturbation on a finite time–interval we show that the reduced system on the center manifold can be brought into the same form. This leads to applications in elliptic problems with localized perturbations, e.g. fluid flow over a bump.

Part II is dedicated to applications of the above theory to elliptic variational problems in cylindrical domains. In Chapter 8 we show that elliptic variational problems in cylinders admit a natural Lagrangian formulation, where the Lagrange function is obtained by simply integrating the density over the cross–section rather than over the whole cylinder. Moreover, the ellipticity condition shows that this Lagrangian problem can always be converted into an associated Hamiltonian problem by the generalized Legendre transformation. Now, the reduction onto a center manifold yields a reduced Hamiltonian system. The question is whether the reduced system on the center manifold again can be understood as the Euler–Lagrange equations of a reduced variational problem. But this can be answered affirmatively if condition (1.4) holds. This provides the first rigorous reduction procedure for elliptic variational problems in cylindrical domains to a finite–dimensional variational problem.

As applications we treat several elliptic problems with increasing complexity. Most of them were already discussed before, however the Hamiltonian structure was not exploited then. The first is just an introductory example to show the basic ideas without involving too much technicalities (Section 8.2). Then we study a nonautonomous problem, describing steady internal waves in a channel in the presence of a small localized perturbation (bump) at the bottom of the channel. The previous results in [Mi86b, Mi88b] can now be obtained much more easily and more completely.

In Chapter 9 steady surface waves under the influence of capillarity and gravity are investigated. In particular, we discuss the question of solitary waves for small Bond number and show that no multi–solitons can bifurcate at Froude number equal to 1.

The fourth example comes from elasticity and considers the deformation of a two–dimensional strip under tensile loading. For certain materials there exists a critical load under which the strip starts to deviate from the homogeneous deformation and exhibits localized necks. We establish the corresponding elliptic variational problem and find the associated reduced variational problem on the four–dimensional center manifold. Thus, the existence of neck solutions can be proved.

The last example (Ch. 11) deals with the so–called nonlinear Saint–Venant problem. In fact, most of the present work was developed in order to handle this problem. It is concerned with static elastic deformations of a very long beam loaded only at its ends. The Euler–Lagange equations are the equilibrium equations of nonlinear elasticity, a strongly elliptic system of three elliptic partial differential equations in three space dimensions. It is shown in [Mi88c] that, for this system, a twelve–dimensional center manifold exists which contains solutions corresponding exactly to the solutions of the classical rod equations given in [Ki59, An72, KMS88]. Moreover, the invariance of the problem under rigid–body transformations is now interpreted as the invariance under the action of the (six–dimensional) Lie group of all Euclidian transformations in $I\!\!R^3$. The undeformed straight beam is a relative equilibrium, since the cross–section moves linearly with the axial variable.

For hyperelastic materials the equilibrium equations are variational and the functional is the stored–energy function $W_{\text{beam}}(\nabla_{(y,t)}u)$. Applying the Lagrangian reduction procedure we are able to construct a reduced variational problem on the tangent bundle of G, giving rise to a hyperelastic rod model with a reduced energy functional for the rod W_{rod} which is deduced from W_{beam} in a mathematically rigorous way. Moreover, we show that, on the quadratic level, a natural reduction can be achieved, i.e. W_{rod} can be chosen such that the energy of a solution calculated in the rod model gives the same value, up to terms of third order in the strains, as the true energy $\int_{\Sigma} W_{\text{beam}} dy$ of the associated beam solution.

Part I

Hamiltonian and Lagrangian theory

Chapter 2

Notations and basic facts on center manifolds

We consider a possibly *infinite–dimensional manifold* $\mathcal{X}$ which is modelled over a reflexive Banach space X ([We71, La62]). As our theory is local with respect to some base point we will often identify $\mathcal{X}$ with X by a local coordinate system. As a general reference for infinite–dimensional Hamiltonian systems we use [CM74, AM78, Ma81] and [HM83, Ch.5.3]. Throughout the whole work we try to be as self–contained as possible. Therefore we avoid the extensive use of special notations and facts in the calculus on manifolds. This should enable readers, who are not familiar with this field, to follow the analysis. However, at some points this has to be payed with less consistent notations or with less elegant proofs.

As usual the *tangent* and the *cotangent bundles* $T\mathcal{X}$ and $T^*\mathcal{X}$ are defined as the unions of the local tangent and cotangent spaces $(x, T_x\mathcal{X})$ and $(x, T_x^*\mathcal{X})$, respectively. Hence, $T\mathcal{X}$ and $T^*\mathcal{X}$ are locally isomorphic to $X \times X$ and $X \times X^*$, respectively, where X^* denotes the dual space of X consisting of all continuous linear forms on X with the natural contraction written as $\langle \cdot, \cdot \rangle_X : X^* \times X \to \mathbb{R}$.

All manifolds, functions, mappings, and bundles over manifolds will only be assumed to have a finite order of differentiability, which is sufficiently high to perform the calculations. A theory of C^k–manifolds is, for instance, developed in [La62]. We cannot work in the C^∞–setting as the center manifold, being our main interest, is in general not a C^∞–manifold (see [CH82]). We will not specify the order of differentiability explicitly in each point. From the context it will be clear what regularity is needed in a certain step. For instance, on a C^k–manifold, being defined to have charts ϕ_i such that the compositions $\phi_i \circ \phi_j^{-1}$ are in C^k, we may define only C^k–functions, the tangent space is a C^{k-1}–manifold, and so on.

Throughout the whole work it suffices to consider C^6–manifolds. Starting with a

C^k–manifold we can define a C^k–Hamiltonian and a symplectic form in C^{k-1}. Thus the Hamiltonian vector field will also be in C^{k-1}. Then typically the center manifold will be in C^{k-r-1} with $r = 0$ in the finite–dimensional case or for semilinear partial differential equations. For quasilinear systems only $r = 1$ is known [Mi88b]. Now the reduced Hamiltonian will be C^{k-r-1} and the reduced symplectic structure C^{k-r-2}. Using Darboux's Theorem we will loose again one order of differentiation since the canonical coordinates are only as smooth as the symplectic structure. Hence, in canonical coordinates the reduced Hamiltonian will be a C^{k-r-2}–function and the associated vector field is in C^{k-r-3}. When Lagrangian problems are considered we need one additional order of differentiation for each involved Legendre transform.

A dense subset $\mathcal{Y} \subset \mathcal{X}$ is called a *manifold domain* in $\mathcal{X}$ if $\mathcal{Y}$ is a manifold in itself, modelled over the Banach space Y being continuously and densely included in X, if the inclusion map $i : \mathcal{Y} \to \mathcal{X}$ is smooth, and if there are charts $\phi : U \subset \mathcal{X} \to X$ such that $\psi = \phi|_{U \cap \mathcal{Y}}$ is a chart $\psi : U \cap \mathcal{Y} \to Y$. Note that $\mathcal{Y}$ is not a submanifold of $\mathcal{X}$ but is only contained as a set, since the topologies of X and Y are different. This generalizes the linear case where Y is just the domain of some relevant linear operator acting on X. A typical manifold example is given by

$$\mathcal{X} = \{\, u \in L_2(\Omega) \ : \ \int_\Omega u^2 \, dy = 1 \,\} \quad \text{and} \quad \mathcal{Y} = \{\, u \in \mathcal{X} \ : \ u \in H^1(\Omega) \,\}$$

where Ω is a domain in $\mathbb{R}^n$ and $H^1(\Omega)$ the Sobolev space of functions with square integrable gradient.

A mapping $v : \mathcal{Y} \to T\mathcal{X}$ with $v_x = v(x) \in T_x\mathcal{X}$ is called a *vector field* on $\mathcal{X}$ with domain $\mathcal{Y}$. We simply say v is a vector field on $\mathcal{X}$ if $\mathcal{Y} = \mathcal{X}$. A vector field defines a differential equation

$$\dot{x} = v(x), \tag{2.1}$$

and a curve $c : (a, b) \subset \mathbb{R} \to \mathcal{X}$ is called a solution of (2.1) if $c : (a, b) \to \mathcal{Y}$ is continuous, $c : (a, b) \to \mathcal{X}$ is differentiable with $\dot{c}(t) = v(c(t))$ for all $t \in (a, b)$.

The directional derivative of a function $f : \mathcal{Y} \to \mathbb{R}$ with respect to the vector field v is denoted by $Df[v]$ and is again a function. The *Lie bracket* $[v, w]$ of two vector fields v and w is the unique vector field defined by the relation

$$Df[[v, w]] = D(Df[w])[v] - D(Df[v])[w]$$

for every function f. (In local coordinates: $[v, w](x) = Dw(x)[v(x)] - Dv(x)[w(x)]$.) A *one–form* α on $\mathcal{X}$ with domain $\mathcal{Y}$ is defined to be a mapping $\alpha : \mathcal{Y} \to T^*\mathcal{X}$ with $\alpha_x \in T_x^*\mathcal{X}$. The contraction of a one–form and a vector field is a real–valued function denoted by $\alpha(v)$, where $\alpha(v)_x = \langle \alpha_x, v_x \rangle_{T_x\mathcal{X}}$. A *two–form* ω on $\mathcal{X}$ is a smooth family ω_x, $x \in \mathcal{X}$, of skew–symmetric bilinear mappings $\omega_x : T_x\mathcal{X} \times T_x\mathcal{X} \to \mathbb{R}$, i.e. $\omega(v_1, v_2) = -\omega(v_2, v_1)$ and $\omega_x(v_1, v_2)$ is linear in v_1 and in v_2.

The *exterior derivative* **d** may be defined for functions, one–forms, and for two–forms as follows:

a) Let f be a function on $\mathcal{X}$, then $\mathbf{d}f(v) = Df[v]$ defines the one–form $\mathbf{d}f$ on $\mathcal{X}$.

b) Let α be a one–form on $\mathcal{X}$, then

$$\mathbf{d}\alpha(v, w) = D(\alpha(w))[v] - D(\alpha(v))[w] - \alpha([v, w])$$

defines the two–form $\mathbf{d}\alpha$ on $\mathcal{X}$.

c) A two–form ω has zero exterior derivative, $\mathbf{d}\omega = 0$, if and only if

$$D(\omega(v_1, v_2))[v_3] + D(\omega(v_2, v_3))[v_1] + D(\omega(v_3, v_1))[v_2]$$
$$-\omega([v_1, v_2], v_3) - \omega([v_2, v_3], v_1) - \omega([v_3, v_1], v_2) \quad = \quad 0$$

for all vector fields v_1, v_2, and v_3.

Note that differential forms on a C^k–manifold are C^{k-1} and their derivatives are C^{k-2}. Moreover, $\mathbf{dd}f = 0$ and $\mathbf{dd}\alpha = 0$ for all functions f and one–forms α.

A one– or two–form β is called *closed* if $\mathbf{d}\beta = 0$, and it is called *exact* if $\beta = \mathbf{d}\gamma$ for a function or one–form γ. The Poincaré Lemma (see Chapter 5) just states that every closed one– or two–form is locally exact.

Each two–form ω defines via $\omega_x(v, w) = \langle \Omega_x v, w \rangle_{T_x\mathcal{X}}$ a family of bounded linear operators $\Omega_x : T_x\mathcal{X} \to T_x^*\mathcal{X}$. The skew–symmetry of ω implies $\Omega_x^* = -\Omega_x$ (here the reflexivity of X is essential [We71]). We say that ω is *weakly non–degenerate* in x if $\Omega_x v = 0$ has only the solution $v = 0$. If moreover the inverse of Ω_x, $J_x = \Omega_x^{-1}$, is a bounded linear operator from $T_x^*\mathcal{X}$ into $T_x\mathcal{X}$, then ω is called *strongly non–degenerate*. Furtheron we always assume strong non–degeneracy. However, in our applications we do reductions onto finite–dimensional submanifolds where weak and strong non–degeneracy are the same. Thus, most results can be generalized to weak nondegeneracy. We will omit the terms "strong" or "strongly" subsequently.

A two–form ω on $\mathcal{X}$ is called a *symplectic form* or a *symplectic structure* on $\mathcal{X}$, if it is closed ($\mathbf{d}\omega = 0$) and if ω_x is (strongly) non–degenerate for all $x \in \mathcal{X}$. A pair $(\mathcal{X}, \omega)$ with ω being a symplectic form is called a *symplectic manifold*. The classical example is $\mathcal{X}$ being $Z \times Z^*$, where Z is some reflexive Banach space. With $(u_i, w_i) \in Z \times Z^* = T_x\mathcal{X}$, the canonical symplectic form ω_{can} is given by

$$\omega_{can}((u_1, w_1), (u_2, w_2)) = \langle w_2, u_1 \rangle_Z - \langle w_1, u_2 \rangle_Z \tag{2.2}$$

In this case we have $\Omega(u, w) = (-w, u)$ and $J(w, u) = (u, -w)$. In [We71, Cor.5.2] it is shown that every constant symplectic form in a Hilbert space can be transformed into a canonical one.

If ω is a constant symplectic form on a linear space X (this applies to $T_x\mathcal{X}$ for every fixed x) then two subspaces Y and Z are called orthogonal if $\omega(y, z) = 0$ for every

$(y, z) \in Y \times Z$. The *orthogonal complement* $Y^\perp$ of Y with respect to ω is given by $\{ x \in X : \omega(y, x) = 0 \text{ for all } y \in Y \}$. A subspace Y is called isotropic if $Y \subset Y^\perp$ and it is called symplectic if the restriction of ω to Y is non–degenerate. A *submanifold* $\mathcal{M}$ of $\mathcal{X}$ is called *symplectic* or *isotropic* if for every $x \in \mathcal{M}$ the tangent space $T_x\mathcal{M}$ is a symplectic or isotropic subspace of $T_x\mathcal{X}$. Note that the restriction $\tilde{\omega}$ of ω onto $T\mathcal{M}$ is again a closed two–form; hence $(\mathcal{M}, \tilde{\omega})$ forms a symplectic manifold in itself. For instance, the center manifolds studied below are shown to be always symplectic submanifolds.

For a smooth function f on $\mathcal{X}$ with domain $\mathcal{Y}$ we define the differential $\mathbf{d}f$ as a one–form with domain

$$\mathcal{D}_f = \{ x \in \mathcal{X} : \mathbf{d}f_x \in T_x^*\mathcal{Y} \text{ extends to } \mathbf{d}f_x \in T_x^*\mathcal{X} \}$$

The extension $\mathbf{d}f(x) \in T_x^*\mathcal{X}$ is uniquely defined as Y is dense in X. In all our applications $\mathcal{D}_f$ is again a manifold domain. For linear systems with $H = \frac{1}{2}\langle Ax, x \rangle$, where $A : D(A) \subset X \to X^*$, the sets $\mathcal{Y}$ and $\mathcal{D}_H$ are given by $D(A^{1/2})$ and $D(A)$, respectively. For a function $H : \mathcal{Y} \to \mathbb{R}$ on a symplectic manifold $(\mathcal{X}, \omega)$ we obtain the *Hamiltonian vector field* X_H on $\mathcal{X}$ with domain $\mathcal{D}_H$ via

$$X_H(x) = J_x \mathbf{d}H_x \qquad \text{or} \qquad \omega(X_H, v) = \mathbf{d}H(v)$$

for every vector field v on $\mathcal{X}$. The differential equation

$$\dot{x} = X_H(x) = J_x \mathbf{d}H_x \tag{2.3}$$

is called the Hamiltonian system generated by the Hamiltonian H.

Using the operator J the associated Poisson bracket $\{ \cdot, \cdot \}$ is defined by

$$\{F, G\} = \mathbf{d}F(J\mathbf{d}G). \tag{2.4}$$

Note that the relation $\{F, G\} = \omega(X_F, X_G)$ holds. The following are the well–known properties of Poisson brackets (cf. [AM78]). For all functions F, G, and H we have

$$
\begin{array}{lll}
\text{bilinearity} & \{F + G, H\} = \{F, H\} + \{G, H\}, & \\
\text{skew–symmetry} & \{F, G\} = -\{G, F\}, & \\
\text{Leibniz's rule} & \{F, GH\} = \{F, G\}H + \{F, H\}G, & \\
\text{Jacobi's identity} & \{F, \{G, H\}\} + \{G, \{H, F\}\} + \{H, \{F, G\}\} = 0. &
\end{array} \tag{2.5}
$$

Any bracket $\{ \cdot, \cdot \}$ satisfying (2.5) is called Poisson structure on $\mathcal{X}$, and a pair $(\mathcal{X}, \{ \cdot, \cdot \})$ is then called a Poisson manifold. Every Poisson bracket is given through (2.4) with a smooth family $J_x : T_x^*\mathcal{X} \to T_x\mathcal{X}$. In general, J need not to be invertible (e.g. $J = 0$ defines a trivial Poisson structure), but if it has a bounded inverse $J_x^{-1} = \Omega_x$, then Ω defines a symplectic structure. Every symplectic manifold can also be considered as a Poisson manifold by using the Poisson bracket associated to ω.

Given a Poisson manifold and a Hamiltonian H, the differential equation (2.3), with J satisfying (2.5), is now called a *generalized Hamiltonian system*. Often it is written in the form

$$\dot{F} = \{F, H\},$$

where F is an arbitrary function on $\mathcal{X}$.

We will not consider existence questions of solutions for the differential equation (2.3). In fact, for the elliptic problems treated below the initial value problem is generally not solvable. For hyperbolic systems we refer to the extensive literatur (e.g. [HKM77]). Our point of view starts with a given *submanifold* $\mathcal{M}$ of $\mathcal{X}$ which is *invariant* under the flow, i.e. $X_H(x) \in T_x\mathcal{M}$ for all $x \in \mathcal{M}$ and every solution starting in $\mathcal{M}$ remains there. The question is whether this submanifold is symplectic. If this is the case we further ask whether the flow on $\mathcal{M}$ coincides with the one generated by the reduced symplectic structure $\widetilde{\omega}$ and the restriction $\widetilde{H}$ of H onto $\mathcal{M}$. For center manifolds these questions will be answered affirmative.

Although in the present work this submanifold will always be a finite–dimensional center manifold we remark that many of the reduction results for the symplectic and Poisson structures can be carried over to much more general situations. However, because of lack of space we only treat the center manifold case.

Center manifolds are very useful tools when the behavior of the system close to an equilibrium is studied. Let the system be described by the differential equation

$$\frac{d}{dt}x = Kx + f(t, \lambda, x), \tag{2.6}$$

where $\lambda \in \mathbb{R}^n$ is a parameter and $f(t, \lambda_0, x) = \mathcal{O}(\|x\|^2)$, i.e. $x = 0$ is an equilibrium for $\lambda = \lambda_0$. We will not assume that the system has an evolutionary property (well–posedness of the initial value problem); thus also elliptic problems as discussed in Chapter 8 can be handled. We are interested in solutions staying close to $x = 0$ for *all* $t \in \mathbb{R}$. They will lie on the center manifold.

First we study the spectral properties of $K : D(K) \to X$. The linear operator K is assumed to be closed, and we consider $D(K)$ as a Banach space equipped with the graph norm. The main hypotheses are:

(C0) The Banach space X splits into two closed subspaces $X_1 \oplus X_2$ which are K–invariant, i.e. $K_j = K|_{X_j} : D(K) \cap X_j \to X_j$ for $j = 1, 2$.

With this assumption the equation (2.6) can be rewritten as a linearly decoupled system

$$\begin{aligned} \dot{x}_1 &= K_1 x_1 + f_1(t, \lambda, x_1 + x_2), \\ \dot{x}_2 &= K_2 x_2 + f_2(t, \lambda, x_1 + x_2). \end{aligned} \tag{2.7}$$

(C1) The spectrum of the linear operator K_1 is contained in the imaginary axis, and K_1 is the generator of a strongly continuous group $(e^{K_1 t})_{t \in \mathbb{R}}$ satisfying $\|e^{K_1 t}\| \le C(1+|t|)^m$ for some $C, m > 0$. (X_1 is called the *center space*).

(C2) The imaginary axis lies in the resolvent set of K_2 and

$$\|(K_2 - i\xi)^{-1}\| \le \frac{C}{1+|\xi|}, \quad \xi \in \mathbb{R}$$

for some $C > 0$.

(C3) There exist $k \in \mathbb{N}$ and a neighborhood $U \subset D(K)$ of 0 and a neighborhood $\Lambda \subset \mathbb{R}^n$ of λ_0 such that $f = f(t, \lambda, x) \in C^{k+1}_{b,unif}(\mathbb{R} \times \Lambda \times U, X)$, $f(t, \lambda_0, 0) = 0$, and $D_x f(t, \lambda_0, 0) = 0$ for all $t \in \mathbb{R}$.

Theorem 2.1

If X is a Hilbert space and system (2.6) satisfies **(C0)** $-$ **(C3)** *then there exist neighborhoods $\tilde{U}_1 \subset U \cap X_1$, $\tilde{U}_2 \subset U \cap X_2$ of 0 and $\tilde{\Lambda} \subset \Lambda$ of λ_0 and a reduction function*

$$h = h(t, \lambda, x_1) \in C^k_{b,unif}(\mathbb{R} \times \tilde{\Lambda} \times \tilde{U}_1, \tilde{U}_2)$$

with $h(t, \lambda_0, 0) = 0$ and $D_{x_1} h(t, \lambda_0, 0) = 0$. The graph of h

$$M_C^\lambda = \{ (t, x_1 + h(t, \lambda, x_1)) \in \mathbb{R} \times X : (t, x_1) \in \mathbb{R} \times \tilde{U}_1 \}$$

is a center manifold for (2.7). This means

a) *M_C^λ is a locally invariant manifold of (2.7), i.e. through every point in M_C^λ there is a solution which stays in M_C^λ as long as it remains in $\tilde{U}_1 \times \tilde{U}_2$.*

b) *Every small bounded solution $x : \mathbb{R} \to D(K)$ which satisfies $x_j(t) \in \tilde{U}_j$, $j = 1, 2$ for all $t \in \mathbb{R}$ lies completely in M_C^λ.*

c) *Every solution $\tilde{x}_1 : (t_1, t_2) \to \tilde{U}_1$ of the reduced equation*

$$\dot{x}_1 = K_1 x_1 + \tilde{f}_1(t, \lambda, x_1) \quad \text{with } \tilde{f}_1(t, \lambda, x_1) = f_1(t, \lambda, x_1 + h(t, \lambda, x_1)) \quad (2.8)$$

leads via $x(t) = x_1(t) + h(t, \lambda, x_1(t))$, $t \in (t_1, t_2)$, to a solution of the full problem (2.7).

d) *If (2.7) is invariant under an isometry $T = T_1 \times T_2$, i.e. $f_j(t, \lambda, T_1 x_1 + T_2 x_2) = T_j f(t, \lambda, x_1 + x_1)$ and $K_j T_j = T_j K_j$, then the reduced vector field $\tilde{f}_1$ is invariant under T_1, i.e. $\tilde{f}_1(t, \lambda, T_1 x_1) = T_1 \tilde{f}_1(t, \lambda, x_1)$.*

For a complete proof of this theorem see [Mi88b]. Note that this version of the center manifold theorem applies to quasilinear systems as it allows the nonlinearitiy f to loose as much regularity as the linear operator K. The result given here suffices for all the applications we will study in Part II.

The first constructions of center manifolds appeared in [Pl64, Ke67] for ODEs. The relevance for infinite dimensional dynamical systems was first pointed out in [RT71]. The symmetry aspects (part d) of the theorem) are developed in [Ru73]. Nowadays there is a big variety of different version of the theorem available, for instance the cases where X is not necessarily a Hilbert space or where the resolvent estimate of (C2) is weaker, are given in [Ca81, He81, Mi91a, Ki91]. For hyperbolic problems, which include the usual infinite dimensional Hamiltonian systems, it is also possible to construct infinite dimensional center manifolds, see [Sc91, Mi89].

The general method of proof is to transform (2.7) into an equivalent integral equation having the same small bounded solutions. Therefore the nonlinearity f is localized by multiplication with an appropriate cut–off function χ. Let $\tilde{f} = f\chi$ then $\tilde{f}$ can be obtained such that it coincides with f is a small neighborhood and still has a small Lipschitz constant. Now, (2.7) can be written in integral form as

$$
\begin{aligned}
x_1(\tau) &= e^{K_1\tau}\xi + \int_0^\tau e^{K_1(\tau-r)}\tilde{f}_1(s+r,\lambda,x(r))\,dr, \\
x_2(\tau) &= \int_{\mathbf{R}} G_2(\tau-r)\tilde{f}_2(s+r,\lambda,x(r))\,dr.
\end{aligned}
\tag{2.9}
$$

Here we have complemented (2.7) with the initial condition $x_1(s) = \xi$ and introduced the shifted time variable $\tau = t - s$. The Green's function G_2 is obtained from the resolvent of K_2 along the imaginary axis by Fourier transform $G(t) = (1/2\pi)\int_{\mathbf{R}} e^{its}(K_2 - is)^{-1}ds$. From (C2) we obtain the estimate $\|G(t)\|_{X_1 \to D(K_1)} \leq Ce^{-\alpha|t|}/|t|$, $t \neq 0$, for some $\alpha > 0$.

Using the contraction mapping principle it is now straight forward to obtain a unique solution $x = Y(\cdot\,; s, \lambda, \xi)$ in a space of exponentially weighted functions. The reduction function is then defined by $h(s, \lambda, \xi) = Y(0; s, \lambda, \xi)$. The proof of the C^k–regularity is more involved and uses the fiber contraction principle on a scale of Banach spaces. From the point of applications it is not neccessary to go into details here. We only give a method to calculate the Taylor expansion of the reduction function directly. For simplicity we omit the parameter λ which could always be considered as a component of x_1. To find the Taylor expansion for h it is more efficient to insert the reduction function into (2.7). We obtain the invariance condition

$$
\frac{\partial}{\partial t}h + D_{x_1}h[K_1x_1 + f_1(t,x_1+h)] = \frac{d}{dt}h(t,\lambda,x_1)
$$
$$
= \dot{x}_2 = K_2h + f_2(t,x_1+h).
\tag{2.10}
$$

Inserting $h(t,x_1) = \sum_2^N h_i(t,x_1) + \mathcal{O}(\|x_1\|^{N+1})$, where the functions h_i are homogeneous in x_1 of degree i, we find that we can solve for the h_i successively. In each step we have to solve a linear system of the form

$$\frac{\partial}{\partial t}h_i + D_{x_1}h_i[K_1 x_1] - K_2 h_i = r_i(t, x_1),$$

where r_i depends only on f_1, f_2, and the values of h_j for $j = 2, \ldots, i - 1$. Under the hypotheses (C1)+(C2) this equation is always solvable; in fact the solution can be given explicitly by

$$h_i(t, x_1) = \int_{\mathbf{R}} G_2(t - s) r_i(s, e^{K_1 s}) \, ds. \tag{2.11}$$

This formula is to be expected when doing the differentiation (formally) in integral equation (2.9) (see also [ETB*87, IA91]). Note that the Taylor expansion of the reduction function is unique, although there may exist different reduction functions depending on the choice of the cut–off function χ used above. However, all the possible center manifolds contain the set of all small bounded orbits. If this set fills a whole manifold then the center manifold must be unique.

     In Chapter 4 the Hamiltonian theory and the center manifold theory are brought together in showing that in the autonomous case the reduced equation (2.8) has actually the form

$$\dot{x}_1 = \tilde{J}(x_1)\mathrm{d}\tilde{H}(x_1) \qquad \text{with } \tilde{H}(x_1) = H(x_1 + h(x_1))$$

and is again a (generalized) Hamiltonian system.

     Throughout the work we do not consider parameter dependent systems. However, they can be treated by extending the system appropriately. Let $\lambda \in \mathbb{R}^n$ be a vector-valued parameter and $H = H(x, \lambda)$ a Hamiltonian on $\mathcal{X}$. Using $\mu \in \mathbb{R}^n$ as an auxiliary variable we define $\overline{\mathcal{X}} = \mathcal{X} \times \mathbb{R}^{2n}$, $\overline{H}(x, \lambda, \mu) = H(x, \lambda)$, and

$$\overline{\omega}_{(x,\lambda,\mu)}((v_1, \lambda_1, \mu_1), (v_2, \lambda_2, \mu_2)) = \omega_x(v_1, v_2) + \lambda_1 \cdot \mu_2 - \lambda_2 \cdot \mu_1.$$

Now the theory developed below is applicable to the enlarged system. Obviously, or more explicitly by using the results of Chapter 5, the analysis and the reduced problem will not depend on μ.

Chapter 3

The linear theory

In the linear Hamiltonian theory the manifold $\mathcal{X}$ is the reflexive Banach space X itself. The symplectic form ω is constant, i.e. Ω and $J = \Omega^{-1}$ are independent of $x \in X$. The only remaining condition on Ω is the skew–symmetry: $\Omega^* = -\Omega$. As $x = 0$ is assumed to be a fixed point, the Hamiltonian H satisfies $\mathbf{d}H(0) = 0$; and, to make (2.3) linear, H has to be homogeneous of degree two. Thus we have

$$H(x) = \frac{1}{2}\langle Ax, x \rangle_X$$

where $A : D(A) \subset X \to X^*$ can be assumed to satisfy $A^* = A$. Hence the linear differential equation is

$$\dot{x} = Kx, \quad \text{where} \quad K = JA : D(A) \to X. \tag{3.1}$$

Our main assumption on the linear operator K is the existence of a positive α such that the resolvent $(K - \lambda)^{-1} : X \to D(A)$ exists for all λ in the strip $\{\lambda \in \mathbb{C} : |\mathrm{Re}\,\lambda| < \alpha\}$ except for finitely many points, each of these points having only a finite–dimensional generalized eigenspace. This guarantees that the spectral part corresponding to the imaginary axis is finite–dimensional while the rest of the spectrum is bounded away from the imaginary axis. This is the typical situation for applying the center manifold theory (cf. [Mi88b]). For an example with an infinite–dimensionl center manifold see [Mi89].

In the first section we prove some properties of the spectral projections associated to K. This will readily imply that the center space X_1, being the spectral part on the imaginary axis, and as well the hyperbolic space X_2, being the spectral part off the imaginary axis, are symplectic subspaces of X. Moreover, A is shown to decompose into $A(x_1 + x_2) = A_1 x_1 + A_2 x_2$ where $x = (x_1, x_2) \in X_1 \times X_2$.

The subsequent section is only needed for the Lagrangian theory in Chapter 6 and can be omitted unless the reader is interested in the proof of Lemma 6.3. We give a list of

those cases of linear normal forms which are needed in Section 6.3. For a complete theory
of linear normal forms we refer to [Wi36, BC74, BC77, Br88].

3.1   Spectral projections

To do the spectral theory we complexify the Banach space X in the usual way and denote
the complexification by $X_{\mathbb{C}} = X \oplus iX$. The symplectic form ω is extended to $X_{\mathbb{C}}$ bilinearly.
Hence we have

$$\omega(u + iv, w + ix) = \omega(u, w) - \omega(v, x) + i(\omega(u, x) + \omega(v, w)).$$

Note that this definition deviates from that of [Kl82], but it simplifies the formulas consid-
erably. In the following we always assume Ω, J, and A to be real, hence $\omega(\bar{x}, \bar{y}) = \overline{\omega(x, y)}$.
A subspace $V_{\mathbb{C}}$ of $X_{\mathbb{C}}$ will be called symplectic, if ω restricted to $V_{\mathbb{C}}$ is non-degenerate.
A subspace $V_{\mathbb{C}} \subset X_{\mathbb{C}}$ is called real if $\overline{V}_{\mathbb{C}} = V_{\mathbb{C}}$. Real subspaces $V_{\mathbb{C}}$ can be interpreted as
complexification of the subspace $V_{\mathbb{R}} \subset X$, where $V_{\mathbb{R}} = \operatorname{Re} V_{\mathbb{C}} = \{ v \in V_{\mathbb{C}} : v = \bar{v} \}$.

The *spectral projection* P_{Σ} of a bounded subset Σ of the spectrum of K, $\operatorname{spec}(K)$, is
defined by the Dunford integral

$$P_{\Sigma}x = \frac{1}{2\pi i} \int_{\Gamma_{\Sigma}} (K - \rho)^{-1}x \, d\rho,$$

where Γ_{Σ} is a bounded curve in the resolvent set of K surrounding exactly those points
of $\operatorname{spec}(K)$ which are in Σ. The curve is oriented in mathematically positive direction
(interior on the left). We always assume that Σ and Γ_{Σ} lie in the strip $|\operatorname{Re} \lambda| < \alpha$ where
the resolvent exists except for a finite number of points.

Lemma 3.1
Let Σ be as above and define $-\Sigma = \{ \lambda \in \mathbb{C} : -\lambda \in \Sigma \}$.
* a) We have $\omega(P_{\Sigma}x, y) = \omega(x, P_{-\Sigma}y)$ for all $x, y \in X$.*
* b) Assume $\Sigma = -\Sigma$; then the spaces $X_1^{\Sigma} = P_{\Sigma}X$ and $X_2^{\Sigma} = (I - P_{\Sigma})X$ are each*
symplectic and satisfy the relation $(X_1^{\Sigma})^{\perp} = X_2^{\Sigma}$.

Proof: The main feature of linear Hamiltonian systems is the so-called skew-symmetry
of K with respect to the symplectic structure ω:

$$\omega(Kv, w) = -\omega(v, Kw). \tag{3.2}$$

for all v and w. This follows immediately from $\langle \Omega J A v, w \rangle = \langle A v, w \rangle = \langle v, \Omega J A w \rangle$ and
the skew-symmetry of Ω.

a) For ρ in the resolvent set of K we deduce from (3.2) the relation $\omega((K-\rho)^{-1}x,y) = -\omega(x,(K+\rho)^{-1}y)$. Hence,

$$\omega(P_\Sigma x, y) = \frac{1}{2\pi i} \int_{\Gamma_\Sigma} \omega((K-\rho)^{-1}x,y) \, d\rho = \omega\left(x, -\frac{1}{2\pi i}\int_{\Gamma_\Sigma}(K+\rho)^{-1}y \, d\rho\right).$$

By letting $\sigma = -\rho$ and changing the orientation the result follows.

b) Letting $x = x_1 + x_2 \in X_1^\Sigma \oplus X_2^\Sigma = X$ we obtain, by using part a) with $\Sigma = -\Sigma$,

$$\omega(x_1, y_2) = \omega(P_\Sigma x_1, y_2) = \omega(x_1, P_\Sigma y_2) = \omega(x_1, 0) = 0.$$

This implies

$$\omega(x_1 + x_2, y_1 + y_2) = \omega(x_1, y_1) + \omega(x_2, y_2) \tag{3.3}$$

which means that $\Omega : X \to X^*$ has the block structure $\Omega = \begin{pmatrix} \Omega_1 & 0 \\ 0 & \Omega_2 \end{pmatrix}$ with $\Omega_k : X_k^\Sigma \to (X_k^\Sigma)^*$ for $k = 1, 2$. Thus the inverse $J = \Omega^{-1}$ has also this structure: $J = \begin{pmatrix} J_1 & 0 \\ 0 & J_2 \end{pmatrix}$ with $J_k = \Omega_k^{-1}$. Hence X_k^Σ is symplectic.

Moreover, $(X_1^\Sigma)^\perp = X_2^\Sigma$ follows from (3.3).   □

Note that the spaces X_1^Σ are real subspaces of $X_{\mathcal{C}}$ if Σ is symmetric with respect to the real axis, i.e. $\overline{\Sigma} = \Sigma$. Specializing this result to $\Sigma = \{\lambda \in \mathcal{C} : \lambda \text{ eigenvalue of } K, \text{ Re } \lambda = 0\}$ leads to the *center space* $X_1 = X_1^\Sigma$ and to the main result needed in the subsequent chapters.

Theorem 3.2

Denote by $P_C : X \to X_1$ the spectral projection onto the center space X_1; then $\omega(P_C x, y) = \omega(x, P_C y)$, X_1 and $X_2 = (I - P_C)X$ are symplectic, and $X_2 = X_1^\perp$.

An additional feature of Hamiltonian systems is the fact that not only the symplectic structure decouples but also the Hamiltonian function H.

Theorem 3.3

Let $\Sigma = -\Sigma$ and $x = x_1 + x_2 \in X_1^\Sigma \oplus X_2^\Sigma = X$. Then the Hamiltonian $H = H(x) = \frac{1}{2}\langle Ax, x \rangle$ has the form

$$H(x_1 + x_2) = \frac{1}{2}\langle A_1 x_1, x_1 \rangle + \frac{1}{2}\langle A_2 x_2, x_2 \rangle \tag{3.4}$$

where $A_k = A|_{D(A)\cap X_k^\Sigma}$.

Proof: As P_Σ is a spectral projection we have $KP_\Sigma = P_\Sigma K$ and thus $K(x_1 + x_2) = K_1 x_1 + K_2 x_2$ with $K_k = K|_{D(A)\cap X_k^\Sigma}$. Hence, $A(x_1+x_2) = \Omega K(x_1+x_2) = \Omega_1 K_1 x_1 + \Omega_2 K_2 x_2$, which defines $A_k = \Omega_k K_k$.

Moreover, $\langle Ax_1, x_2 \rangle = \omega(Kx_1, x_2) = \omega(P_\Sigma K_1 x_1, x_2) = \omega(K_1 x_1, P_\Sigma x_2) = \omega(K_1 x_1, 0) = 0$. This completes the proof.   □

Remark: The same splitting result for H holds whenever K and Ω have block structure with respect to the splitting $X = X_1 \times X_2$.

3.2   Linear normal forms

Henceforth we may assume that the underlying space X is finite–dimensional, since Lemma 3.1 and Theorem 3.3 allow us to reduce the problem to any spectral part X_1^Σ which satisfies $\Sigma = -\Sigma = \overline{\Sigma}$.

For each λ in the spectrum of K the *generalized eigenspace*

$$V(\lambda) = \{\, x \in X \;:\; (K - \lambda)^j x = 0 \text{ for some } j \in I\!\!N \,\}$$

is also given by $V(\lambda) = P_\lambda X$, where $P_\lambda = P_{\{\lambda\}}$ is the spectral projection. We need the following result which is also proved in [Kl82, Ch.3.1].

Theorem 3.4
For any λ and μ in $\mathbb{C}$ we have:
   a) If $\lambda + \mu \neq 0$, then $V(\lambda)$ and $V(\mu)$ are orthogonal with respect to ω.
   b) $V(\lambda) \oplus V(-\lambda)$ is a symplectic subspace of $X_{\mathbb{C}}$.
   c) $W(\lambda) = Re(V(\lambda) \oplus V(-\lambda) \oplus V(\overline{\lambda}) \oplus V(-\overline{\lambda}))$ is a symplectic subspace of X.

Proof: The assertions b) and c) follow immediately by applying Lemma 3.1 with $\Sigma = \{\lambda, -\lambda\}$ and $\Sigma = \{\lambda, -\lambda, \overline{\lambda}, -\overline{\lambda}\}$, respectively.

To prove assertion a) we note that, because of $\mu \neq -\lambda$, the spectral projections $P_{-\lambda}$ and P_μ satisfy $P_{-\lambda} P_\mu = 0$. Hence, for all $(v, w) \in V(\lambda) \times V(\mu)$ we have

$$\omega(v, w) = \omega(P_\lambda v, P_\mu w) = \omega(v, P_{-\lambda} P_\mu w) = 0,$$

which is the desired result.   □

As we have seen in Theorem 3.3 the Hamiltonian system decouples completely on the subspaces $W(\lambda)$. Hence, it is sufficient to consider each subspace by itself. Here we restrict ourself to the case that λ lies on the imaginary axis, since this is the only case needed below. However, the provided methods are easily adjusted to fit the general case. Whenever a result is restricted to λ being on the imaginary axis, we write is, $s \in I\!\!R$ instead of λ. E.g. $W(is) = V(is) \oplus V(-is)$ if $s \neq 0$ and $W(0) = V(0)$.

Note that the generalized eigenspace $V(\lambda)$ has a basis consisting of *Jordan chains* $(x_{j,k})_{j=1,\ldots,m(k)}$ with $k = 1,\ldots,l$ and $m(1) + \cdots + m(l) = n = \dim V(\lambda)$. This means that $(K - \lambda)x_{j,k} = x_{j-1,k}$ for $j = 1,\ldots,m(k)$ where $x_{0,k} = 0$.

Lemma 3.5
Let $(x_{j,k})$ be a basis on Jordan chains of $V(\lambda)$. Then there is a unique basis $(y_{j,k})$ of $V(-\lambda)$ such that

$$\omega(x_{j,k}, y_{r,s}) = \delta_{k-s}\delta_{j-r}. \tag{3.5}$$

Furthermore, the relations

$$(K + \lambda)y_{j,k} = -y_{j+1,k} \tag{3.6}$$

hold for all $j = 1,\ldots,m(k)$ and $k = 1,\ldots,l$.

Proof: The construction of $y_{j,k}$ is as follows. Let $n^{\pm} = \dim V(\pm\lambda)$ and consider the mapping $T : V(-\lambda) \to I\!\!R^{n^+}; y \to (\omega(x_{1,1}, y),\ldots,\omega(x_{m(l),l}, y))$. With Theorem 3.4a+b we conclude that T is injective. This implies that $n^- \geq n^+$. Interchanging λ and $-\lambda$ yields $n^+ \geq n^-$. However, $n^+ = n^-$ implies that T is even one-to-one; and hence, keeping r and s fixed, there is exactly one $y_{r,s}$ such that (3.5) holds for all j and k.

Employing the skew–symmetry of K, (3.2), gives us

$$\omega(x_{j-1,k}, y_{r,s}) = \omega((K - \lambda)x_{j,k}, y_{r,s}) = -\omega(x_{j,k}, (K + \lambda)y_{r,s}). \tag{3.7}$$

Now the uniqueness of $y_{r,s}$ yields (3.6).   □

Note that the vectors $z_{j,k} = (-1)^{j-1}y_{m(k)+1-j,k}$ again form a basis of Jordan chains of $V(-\lambda)$. Moreover, the subspaces $V^k(\lambda) \subset V(\lambda) \oplus V(-\lambda)$ defined by

$$V^k(\lambda) = \mathrm{span}\{x_{1,k},\ldots,x_{m(k),k}, z_{1,k},\ldots,z_{m(k),k}\}$$

are symplectic and invariant under the application of K. Hence, Theorem 3.3 shows that the problem can be further reduced by considering each $V^k(\lambda)$ separately. As $V^k(\lambda)$ is not necessarily a real subspace of $X_{\mathbb{C}}$ we define the real subspaces $W^k(\lambda)_{\mathbb{C}} = V^k(\lambda) \oplus \overline{V^k(\lambda)}$. Now $W^k(\lambda)_{\mathbb{C}}$ maximally consists of four different Jordan chains of the same length $m(k)$. In the following we drop the sub– and superscript k, since we can consider each particular k separately.

The aim of the normal form theory is to find canonical coordinates (q, p) in $W(\lambda)_{\boldsymbol{R}}$ such that the restriction of the Hamiltonian H onto $W(\lambda)_{\boldsymbol{R}}$ takes a simple normalized form depending only on λ, the number and the length m of the Jordan chains. In this sense all linear Hamiltonian systems having the same spectral properties are equivalent. Hence, it suffices for certain facts to consider only systems in normal form. In Chapter 6 we will use the normal forms developed here to show that most Hamiltonian systems can be transformed into a Lagrangian system.

Restricting our attention now to $\lambda = is$ with $s \neq 0$, we have to distinguish two cases: $W(is)$ may consist of two or four Jordan chains. There are only two Jordan chains, if $z_j = \alpha \bar{x}_j$ for some fixed $\alpha \in \mathbb{C}$. This is equivalent to the condition $\omega(x_1, \bar{x}_m) \neq 0$. In each of these cases we obtain a different normal form for the Hamiltonian system on $W(\lambda)$.

Case 1: $s \neq 0$, $\omega(x_1, \bar{x}_m) \neq 0$

From (3.7) we derive

$$0 \neq \beta = \omega(x_1, \bar{x}_m) = (-1)^{m-1}\omega(x_m, \bar{x}_1) = (-1)^m \bar{\beta}. \tag{3.8}$$

Hence, β has to be real if m is even and it has to be purely imaginary for odd m. Without loss of generality we may assume $|\beta| = 2$ after an appropriate scaling of x_j.

Case 1.1: m is even.

The general $x \in W(is)_{\mathbf{R}}$ has the representation

$$x = \sum_{j=1}^{m} r_j \mathrm{Re}\, x_j + t_j \mathrm{Im}\, x_j.$$

Using the Jordan chain property of (x_j) yields $Kx = \tilde{x} = \sum_{j=1}^{m} \tilde{r}_j \mathrm{Re}\, x_j + \tilde{t}_j \mathrm{Re}\, x_j$ with

$$\tilde{r}_j = st_j + r_{j+1}, \qquad \tilde{t}_j = -sr_j + t_{j+1}$$

where $r_{m+1} = t_{m+1} = 0$. From $\omega(x_j, x_k) = 0$ and $\omega(x_j, \bar{x}_k) = \pm 2\delta_{m+1-j-k}$ we obtain

$$\omega(\mathrm{Re}\, x_j, \mathrm{Re}\, x_k) = \omega(\mathrm{Im}\, x_j, \mathrm{Im}\, x_k) = \pm(-1)^j \delta_{m+1-j-k}$$

and $\omega(\mathrm{Re}\, x_j, \mathrm{Im}\, x_k) = 0$ for all j and k. Hence, the symplectic structure reduces to

$$\omega(x, \hat{x}) = \pm \sum_{j=1}^{m} (-1)^{j-1}(r_j \hat{r}_{m+1-j} + t_j \hat{t}_{m+1-j})$$

where $\hat{x} = \sum_{j=1}^{m} \hat{r}_j \mathrm{Re}\, x_j + \hat{t}_j \mathrm{Im}\, x_j$. Now letting $q_{2j-1} = r_{2j-1}$, $q_{2j} = t_{2j-1}$, $p_{2j-1} = r_{m+2-2j}$, and $p_{2j} = t_{m+2-2j}$, for $j = 1, \ldots, m/2$, we have found canonical coordinates (q, p). In these coordinates K takes the form

$$
\tilde{K}\begin{pmatrix} q \\ p \end{pmatrix} = \pm
\left(
\begin{array}{ccccc|cccc}
S & & & & & & & & I \\
 & S & & & & & & I & \\
 & & \ddots & & & & \iddots & & \\
 & & & S & I & & & & \\
\hline
 & & & 0 & S & & & & \\
 & & 0 & I & & S & & & \\
 & \iddots & \ddots & & & & \ddots & & \\
0 & I & & & & & & & S
\end{array}
\right)
\begin{pmatrix} q \\ p \end{pmatrix},
$$

where $S = \left(\begin{smallmatrix} 0 & s \\ -s & 0 \end{smallmatrix} \right)$ and $I = \left(\begin{smallmatrix} 1 & 0 \\ 0 & 1 \end{smallmatrix} \right)$. (Places without an entry have to be filled with zeros.)
Thus, the operator $A = \Omega K = \left(\begin{smallmatrix} 0 & -I \\ I & 0 \end{smallmatrix} \right) K$ can be reconstructed completely in this basis. In
particular, the Hamiltonian reads

$$
\pm 2 H(q,p) \;=\; \sum_{j=1}^{m/2} (p_{2j-1} p_{m+1-2j} + p_{2j} p_{m+2-2j}) + 2s \sum_{j=1}^{m/2} (q_{2j} p_{2j-1} - q_{2j-1} p_{2j})
$$

$$
- \sum_{j=1}^{m/2-1} (q_{2j+1} q_{m+1-2j} + q_{2j+2} q_{m+2-2j}).
$$

This is one possible normal form for the case 1.1.

Case 1.2: m is odd.

We have the same representation of x and $\tilde{x} = Kx$ in terms of $\operatorname{Re} x_j$ and $\operatorname{Im} x_j$ as above.
Only the symplectic form differs:

$$
\omega(x, \hat{x}) = \pm \sum_{j=1}^{m} (-1)^{j-1} (r_j \hat{t}_{m+1-j} - t_j \hat{r}_{m+1-j}).
$$

Canonical coordinates are obtained by $q_j = r_j$ and $p_j = (-1)^{j-1} t_{m+1-j}$. Now, K takes
the form

$$
\tilde{K}\begin{pmatrix} q \\ p \end{pmatrix} = \pm
\left(
\begin{array}{ccccc|ccccc}
0 & 1 & & & & & & & & +s \\
 & 0 & \ddots & & & & & & -s & \\
 & & \ddots & 1 & & & & \iddots & & \\
 & & & 0 & +s & & & & & \\
\hline
 & & & -s & 0 & & & & & \\
 & & +s & & & -1 & 0 & & & \\
 & \iddots & & & & & \ddots & \ddots & & \\
-s & & & & & & & & -1 & 0
\end{array}
\right)
\begin{pmatrix} q \\ p \end{pmatrix}.
$$

The normal form of the Hamiltonian is given by

$$\pm 2H(q,p) = s \sum_{j=1}^{m}(-1)^{j-1}(q_j q_{m+1-j} + p_j p_{m+1-j}) + 2\sum_{j=1}^{m-1} p_j q_{j+1}.$$

Case 2: $s \neq 0$, $\omega(x_1, \bar{x}_m) = 0$

Now we have four Jordan chains and $x \in W(is)_R$ has the representation

$$x = \sum_{j=1}^{m} r_j \mathrm{Re}\, x_j + t_j \mathrm{Im}\, x_j + u_j \mathrm{Re}\, y_j + v_j \mathrm{Im}\, y_j$$

where $Kx_j = isx_j + x_{j-1}$, $Ky_j = -isy_j - y_{j-1}$, and $\omega(x_j, y_k) = 2\delta_{m+1-j-k}$. The operator K has the form $\tilde{x} = Kx$ where

$$\tilde{r}_j = st_j + r_{j+1}, \quad \tilde{t}_j = -sr_j + t_{j+1}, \quad \tilde{u}_j = -sv_j - u_{j+1}, \quad \tilde{v}_j = +su_j - v_{j+1}.$$

The symplectic form reduces to

$$\omega(x,\hat{x}) = \sum_{j=1}^{m}(r_j\widehat{u}_{m+1-j} - t_j\widehat{v}_{m+1-j} - u_j\widehat{r}_{m+1-j} + v_j\widehat{t}_{m+1-j}).$$

The canonical coordinates $q = (r_1, t_1, \ldots, r_m, t_m)$ and $p = (u_m, -v_m, u_{m-1}, \ldots, u_1, -v_1)$ transform the operator K into

$$\tilde{K}\begin{pmatrix} q \\ p \end{pmatrix} = \left(\begin{array}{cccc|cccc} S & I & & & & & & \\ & S & \ddots & & & & & \\ & & \ddots & I & & & & \\ & & & S & & & & \\ \hline & & & & S & & & \\ & & & & -I & S & & \\ & & & & & \ddots & \ddots & \\ & & & & & & -I & S \end{array}\right)\begin{pmatrix} q \\ p \end{pmatrix}.$$

The normal form of the Hamiltonian is given by

$$H(q,p) = s \sum_{j=1}^{m}(q_{2j}p_{2j-1} - q_{2j-1}p_{2j}) + \sum_{j=1}^{2m-2} q_{j+2}p_j.$$

Thus, the case $s \neq 0$ is finished.

When dealing with $\lambda = 0$ we have the advantage that all Jordan chains can be chosen real. There are again two cases with either one or two Jordan chains.

Case 3: $s = 0$, $\omega(x_1, x_m) \neq 0$.

We have just one Jordan chain (x_j). From (3.8) we immediately conclude that m has to be even. Letting $x = \sum_{j=1}^{m} r_j x_j$ we obtain, in the notations from above, $\tilde{r}_j = r_{j+1}$ and $\omega(x, \hat{x}) = \sum_{j=1}^{m} (-1)^{j-1} r_j \hat{r}_{m+1-j}$. Choosing the canonical coordinates $q_j = r_{2j-1}$ and $p_j = r_{m+2-2j}$ yields

$$
\tilde{K}\begin{pmatrix} q \\ p \end{pmatrix} = \left(
\begin{array}{ccccc|ccc}
 & & & & 1 & & & \\
 & & & 1 & & & & \\
 & & \ddots & & & & & \\
 & 1 & & & & & & \\
\hline
 & & & & & 0 & & \\
 & & & & & 0 & 1 & \\
 & & & & & & \ddots & \\
0 & 1 & & & & & &
\end{array}
\right) \begin{pmatrix} q \\ p \end{pmatrix}
$$

and the Hamiltonian

$$
2H(q, p) = \sum_{j=1}^{n} p_j p_{n+1-j} - \sum_{j=2}^{n} q_j q_{n+2-j}, \quad n = m/2.
$$

Case 4: $s = 0$, $\omega(x_1, x_m) = 0$.

Now there are two Jordan chains (x_j) and (y_j) with $Kx_j = x_{j-1}$, $Ky_j = -y_{j-1}$, and $\omega(x_j, y_k) = \delta_{m+1-j-k}$. Hence, we have the representations $x = \sum_{j=1}^{m} r_j x_j + t_j y_j$, where $\tilde{r}_j = r_{j+1}$, $\tilde{t}_j = -t_{j+1}$, and $\omega(x, \hat{x}) = \sum_{j=1}^{m}(r_j \hat{t}_{m+1-j} - t_j \hat{r}_{m+1-j})$. Possible canonical coordinates are $q_j = r_j$ and $p_j = t_{m+1-j}$, giving

$$
\tilde{K}\begin{pmatrix} q \\ p \end{pmatrix} = \left(
\begin{array}{cccc|cccc}
0 & 1 & & & & & & \\
 & 0 & \ddots & & & & & \\
 & & \ddots & 1 & & & & \\
 & & & 0 & & & & \\
\hline
 & & & & 0 & & & \\
 & & & & -1 & 0 & & \\
 & & & & & \ddots & \ddots & \\
 & & & & & & -1 & 0
\end{array}
\right) \begin{pmatrix} q \\ p \end{pmatrix},
$$

and the Hamiltonian

$$H(q,p) = \sum_{j=1}^{m-1} q_{j+1}p_j.$$

Chapter 4

Hamiltonian flows on center manifolds

Since the center manifold theory is local with respect to an equilibrium point $x = 0$, we furtheron identify the manifold $\mathcal{X}$ with the Banach space X by means of a local coordinate system. Thus, we consider a Hamiltonian system in the space $X = X_1 \oplus X_2$ where the splitting is done in accordance with the construction of the previous chapter using the linearization at $x = 0$:

$$\begin{pmatrix} \dot{x}_1 \\ \dot{x}_2 \end{pmatrix} = J(x) \begin{pmatrix} \mathbf{d}_1 H(x) \\ \mathbf{d}_2 H(x) \end{pmatrix}, \quad J = \begin{pmatrix} J_1 & J_{12} \\ J_{21} & J_2 \end{pmatrix}. \tag{4.1}$$

The notation $\mathbf{d}_k H$ is used such that $\mathbf{d}_k H(x) \in X_k^*$ and $\mathbf{d}H_x(v) = \mathbf{d}_1 H(v_1) + \mathbf{d}_2 H(v_2)$. X_1 is the finite dimensional center space of the equation and X_2 is the possibly infinite dimensional hyperbolic space, i.e. if we rewrite (4.1) in the form

$$\begin{pmatrix} \dot{x}_1 \\ \dot{x}_2 \end{pmatrix} - K \begin{pmatrix} x_1 \\ x_2 \end{pmatrix} = \begin{pmatrix} f_1(x) \\ f_2(x) \end{pmatrix} = \mathcal{O}(\|x\|^2), \tag{4.2}$$

then $K \begin{pmatrix} x_1 \\ x_2 \end{pmatrix} = \begin{pmatrix} K_1 & 0 \\ 0 & K_2 \end{pmatrix} \begin{pmatrix} x_1 \\ x_2 \end{pmatrix}$ with Re spec $K_1 = 0$ and |Re spec $K_2| \geq \alpha > 0$.

Under additional technical assumptions (see e.g. Theorem 2.1) we know that all sufficiently small bounded solutions of (4.1) lie on a finite dimensional center manifold

$$M_C = \{ x_1 + h(x_1) \in X : \|x\| \ll 1 \}$$

where the reduction function $h : U_1 \subset X_1 \rightarrow X_2 \cap D(K)$ satisfies $h(x_1) = \mathcal{O}(\|x_1\|^2)$.

The *reduced equation* describes the flow on the center manifold and is given by

$$\dot{x}_1 = K_1 x_1 + f_1(x_1 + h(x_1)). \tag{4.3}$$

Our aim is to show that (4.3) is a Hamiltonian system when (4.1) is one. Although in the symplectic case this result seems to be well–known, it is difficult to find references; see [Po80, Mo77]. In Section 4.2 we generalize this result to Poisson manifolds, i.e. that J_1 is not necessarily invertible. In fact, in both cases the reduction method is independent of the center manifold context and applies to general submanifolds which are invariant under the flow of a (generalized) Hamiltonian system and satisfy assumption (**A2**) given below. Up to now very few is known about reduction of Poisson structures to submanifolds [MR86].

In the last section we derive the method of flattening of center manifolds by canonical changes of coordinates. As a general result we show that there are always canonical coordinates (q_1, q_2, p_1, p_2) such that the center manifold is given by $(q_2, p_2) \equiv 0$. In these coordinates the reduction is trivial: The reduced Hamiltonian is $\widetilde{H}(q_1, p_1) = H(q_1, 0, p_1, 0)$ and (q_1, p_1) are canonical coordinates on the center manifold. Moreover, we give an effective way to approximate these coordinates, up to any given order of accuracy, by doing successive almost identical canonical transformations. After n steps, the reduced system will have correct terms up to order $2n$. This methods is especially useful for numerical or symbolical calculation; it was found independently by C. Simò [Si90]. Finally we pose the question of analyticity of center manifolds in analytical Hamiltonian systems. Although we present a counter example, there is still some hope that, under natural conditions, analyticity can be obtained.

4.1   Reduction in the symplectic case

For this case we give a short proof using the methodology of differential geometry. A more analytical proof can be obtained by specializing the proof of Theorem 4.6 to the symplectic case.

Theorem 4.1
Assume that the center manifold is a C^2-manifold. Then the reduced equation (4.3) is the Hamiltonian system corresponding to the reduced symplectic structure $\widetilde{\omega} = \omega|_{M_C}$ and and the reduced Hamiltonian $\widetilde{H} = H|_{M_C}$.

Proof: First we note that $\widetilde{\omega}$ is again a closed two–form ($d\widetilde{\omega} = 0$) since restriction and exterior derivative commute. Moreover, in $x = 0$ we have $\widetilde{\omega}_0(v_1, v_2) = \langle \Omega_1(0)v_1, v_2 \rangle$ with $\Omega_1(0)$ being invertible according to Theorem 3.2. Thus, $\widetilde{\omega}$ is non–degenerate, and hence symplectic, in a whole neighborhood of $x = 0$.

The only thing, that remains to be shown, is whether the original vector field X_H, defined on X by $\omega(X_H, v) = dH(v)$, coincides on M_C with the vector field $X_{\widetilde{H}}$ defined by $\widetilde{\omega}(X_{\widetilde{H}}, \widetilde{v}) = d\widetilde{H}(\widetilde{v})$. However, by the invariance of M_C under the flow of (4.1) we know that $X_H|_{M_C} \in TM_C$, hence $\mathbf{d}H(\tilde{v}) = \omega(X_H, \tilde{v}) = \tilde{\omega}(X_H, \tilde{v})$ for all $\tilde{v} \in TM_C$. Since additionally $\mathbf{d}H(\tilde{v}) = \mathbf{d}\tilde{H}(\tilde{v})$ we find $\tilde{\omega}(X_H, \tilde{v}) = \mathbf{d}\tilde{H}(\tilde{v})$ which is exactly the definition of $X_{\tilde{H}}$. Hence, $X_H|_{M_C} = X_{\tilde{H}}$. $\qquad\square$

These arguments are implicitly used in all the work where this result was applied without further explanation. Although this proof is very short and elegant, we prefer the one given for Theorem 4.6. One reason is that the above proof does not show some of the relevant features related to reductions of this kind. Moreover, it is clearly restricted to the symplectic case.

When explicit calculations have to be performed we need to express the reduced symplectic structure in coordinates. According to (4.1) the coordinates $x_1 \in X_1$ are the most simple choice. The full symplectic structure ω is given by

$$\omega_x(v_1 + v_2, w_1 + w_2) = \langle \Omega_1 v_1, w_1 \rangle + \langle \Omega_{12} v_1, w_2 \rangle + \langle \Omega_{21} v_2, w_1 \rangle + \langle \Omega_2 v_2, w_2 \rangle.$$

Note that a vector field v_1 on X_1 (i.e. $v_1(x_1) \in T_{x_1} X_1$) corresponds to the vector field $v = v_1 + Dh(x_1)v_1$ on M_C, since M_C is the graph of h. Hence,

$$
\begin{aligned}
\tilde{\omega}(v_1, w_1) &= \langle \tilde{\Omega}(x_1) v_1, w_1 \rangle \quad \text{with} \\
\tilde{\Omega}(x_1) &= \Omega_1(x_1 + h(x_1)) + Dh^*(x_1)\Omega_{12}(x_1 + h(x_1)) \\
&\quad + \Omega_{21}(x_1 + h(x_1))Dh(x_1) + Dh^*(x_1)\Omega_2(x_1 + h(x_1))Dh(x_1).
\end{aligned}
\tag{4.4}
$$

This shows that even when starting in a canonical situation, i.e. $\Omega = \text{const.}$, the reduction will, in general, lead to a non–canonical structure on M_C, because the center manifold is not flat. However, for practical purposes the following results may be helpful.

Theorem 4.2
Assume that $\mathcal{X} = T^*Q$ for some reflexive Banach space Q and that $\omega = \omega_{can}$ as in (2.2). Furthermore, assume $H(x) = \frac{1}{2}\langle Ax, x \rangle + N(x)$ with $N(x) = \mathcal{O}(\|x\|^n)$, $DN(x) = \mathcal{O}(\|x\|^{n-1})$ for $x \to 0$, for some $n > 2$. Then there is a subspace E of the center space X_1 such that X_1 can be identified with T^*E. Considering $\tilde{\omega}$ and $\tilde{H}$ as functions on T^*E they satisfy the relations $\tilde{\Omega}(x_1) = \begin{pmatrix} 0 & I \\ -I & 0 \end{pmatrix} + \mathcal{O}(\|x_1\|^{2n-4})$ and $\tilde{H}(x_1) = H(x_1 + 0) + \mathcal{O}(\|x_1\|^{2n-2})$ for $x \to 0$.

This observation, with $n = 3$, was already found in [Mo77].

Corollary 4.3
In the case of Theorem 4.2 with $x_1 = (q, p) \in T^*E$, the reduced equation (4.3) has the form

$$\begin{pmatrix} \dot{q} \\ \dot{p} \end{pmatrix} = \tilde{J}(q, p) \begin{pmatrix} \mathbf{d}_q \tilde{H} \\ \mathbf{d}_p \tilde{H} \end{pmatrix} = \begin{pmatrix} \mathbf{d}_p \hat{H}(q, p) \\ -\mathbf{d}_q \hat{H}(q, p) \end{pmatrix} + \mathcal{O}(\|(q, p)\|^{2n-3}),$$

where $\hat{H}(q, p) = H((q, p) + 0)$.

At the end of the next chapter (see Theorem 5.15) we show that we can find coordinates $(\overline{q}, \overline{p})$ such that the symplectic form is the canonical one $(\overline{\Omega} = \begin{pmatrix} 0 & I \\ -I & 0 \end{pmatrix})$ and that $\overline{H}(\overline{q}, \overline{p}) = H(\overline{q}, \overline{p}) + \mathcal{O}(|(\overline{q}, \overline{p})|^{2n-2})$. Hence, to obtain a good approximation it will often be sufficient to find canonical coordinates (q, p) in the tangent space and to expand H in these coordinates.

These results justify many projection methods (or Galerkin approximations) to lowest order. The present theory shows how the correct higher order terms have to be derived. Another, more effective way of getting such an expansion is described in Section 4.3, where canonical transformations for the full system are used to flatten the center manifold.

Proof of Theorem 4.2: The assumptions on H imply that the function $f = (f_1, f_2)$ in (4.2) satisfies $\|f\| = \mathcal{O}(\|x\|^{n-1})$. Hence, taking into acount (2.10) we obtain $\|h(x_1)\| = \mathcal{O}(\|x_1\|^{n-1})$. Using (3.4) gives $\Omega_{12}(x) = \Omega_{21}(x) = 0$, and thus (4.4) yields the result for $\widetilde{\Omega}$.

For $\widetilde{H}$ we proceed by invoking Theorem 3.3, whence

$$
\begin{aligned}
\widetilde{H}(x_1) &= H(x_1 + h(x_1)) = \frac{1}{2}\langle A_1 x_1, x_1 \rangle + \frac{1}{2}\langle A_2 h(x_1), h(x_1) \rangle + N(x_1 + h(x_1)) \\
&= \frac{1}{2}\langle A_1 x_1, x_1 \rangle + \mathcal{O}(\|x_1\|^{2n-2}) + N(x_1) + [N(x_1 + h(x_1)) - N(x_1)] \\
&= H(x_1 + 0) + \mathcal{O}(\|x_1\|^{2n-2}) + \int_0^1 DN(x_1 + \theta h(x_1))[h(x_1)]\, d\theta \\
&= H(x_1 + 0) + \mathcal{O}(\|x_1\|^{2n-2}) + \mathcal{O}(\max_{\theta \in [0,1]} \|x_1 + \theta h(x_1)\|^{n-1}) \|h(x_1)\| \\
&= H(x_1 + 0) + \mathcal{O}(\|x_1\|^{2n-2}).
\end{aligned}
$$

$\square$

Example: To illustrate the reduction procedure we work through the following example in $T^* \mathbb{R}^2 = \mathbb{R}^4$:

$$
H(q, Q, p, P) = \tfrac{1}{2}(p^2 + \nu^2 q^2) + QP + q^2(\alpha Q + \beta P).
$$

We have $x_1 = (q, p)$ and $x_2 = (Q, P)$, and the linear part is already decoupled. Thus, the center manifold can be found in the form $Q = h(q, p)$, $P = g(q, p)$. The invariance condition (2.11) results in

$$
\begin{pmatrix} \dot{Q} \\ \dot{P} \end{pmatrix} = \begin{pmatrix} \partial_q h & \partial_p h \\ \partial_q g & \partial_p g \end{pmatrix} \begin{pmatrix} p \\ -\nu^2 q - 2qh \end{pmatrix} = \begin{pmatrix} h + \alpha q^2 \\ -g - \beta q^2 \end{pmatrix}.
$$

Inserting the expansion $h(q, p) = h_1 q^2 + h_2 qp + h_3 p^2 + h_4 q^3 + \ldots + h_7 p^3 + \mathcal{O}(q^4 + p^4)$ and similarly for g and comparing equal powers leads to

$$
\begin{aligned}
h(q, p) &= -\alpha \nu_4^{-1}(\nu_2 q^2 + 2qp + 2p^2) + \mathcal{O}(q^4 + p^4), \\
g(q, p) &= -\beta \nu_4^{-1}(\nu_2 q^2 - 2qp + 2p^2) + \mathcal{O}(q^4 + p^4),
\end{aligned}
$$

where $\nu_i = 1 + i\nu^2$. Thus the reduced Hamiltonian $\tilde{H}$ reads

$$\tilde{H}(q,p) = \tfrac{1}{2}(p^2 + \nu^2 q^2) + \nu_2\nu_4^{-1}(\alpha\beta\nu_2\nu_4^{-1} - \alpha^2 - \beta^2)q^4 + 2\nu_4^{-1}(\beta^2 - \alpha^2)q^3 p +$$
$$+ 2\nu_4^{-1}(\nu^2\alpha\beta - \alpha^2 - \beta^2)q^2 p^2 + 0qp^3 + 4\nu_4^{-2}\alpha\beta p^4 + \mathcal{O}(q^6 + p^6).$$

However, (q,p) are no longer canonical coordinates on the center manifold, the reduced symplectic form is given by $\tilde{\Omega}(q,p) = \left(\begin{smallmatrix} 0 & \tilde{\sigma} \\ -\tilde{\sigma} & 0 \end{smallmatrix}\right)$ where

$$\tilde{\sigma}(q,p) = 1 - \partial_q h \partial_p g - \partial_p h \partial_q g = 1 + 8\nu_4^{-2}\alpha\beta(2p^2 - \nu_2 q^2) + \mathcal{O}(q^4 + p^4).$$

To obtain a canonical system we use a change of coordinates given by $q = \bar{q} + r(\bar{q},\bar{p})$, $p = \bar{p} + s(\bar{q},\bar{p})$, with $r(\bar{q},\bar{p}) = r_1\bar{q}^3 + r_2\bar{q}^2\bar{p} + r_3\bar{q}\bar{p}^2 + r_4\bar{p}^3 + \mathcal{O}(|(\bar{q},\bar{p})|^5)$ and similarly for s. The transformed symplectic structure is $\bar{\Omega}(\bar{q},\bar{p}) = \left(\begin{smallmatrix} 0 & \bar{\sigma} \\ -\bar{\sigma} & 0 \end{smallmatrix}\right)$ with

$$\bar{\sigma}(\bar{q},\bar{p}) = \tilde{\sigma}(\bar{q}+r, \bar{p}+s)((1+\partial_{\bar{q}} r)(1+\partial_{\bar{p}} s) - \partial_{\bar{p}} r \partial_{\bar{q}} s)$$
$$= 1 + (3r_1 + s_2 - 8\nu_2\nu_4^{-2}\alpha\beta)\bar{q}^2 + 2(r_2 + s_3)\bar{q}\,\bar{p} +$$
$$+ (r_3 + 3s_4 + 16\nu_4^{-2}\alpha\beta)\bar{p}^2 + \mathcal{O}(\bar{q}^4 + \bar{p}^4).$$

By choosing $s_2 = 8\nu_2\nu_4^{-2}\alpha\beta$, $r_3 = -16\nu_4^{-2}\alpha\beta$, and all other $r_i = s_i = 0$, the new Hamiltonian has the form

$$\bar{H}(\bar{q},\bar{p}) = \tilde{H}(\bar{q}+r(\bar{q},\bar{p}), \bar{p}+s(\bar{q},\bar{p})) = \tilde{H}(\bar{q},\bar{p}) + (s_2 + \nu^2 r_3)\bar{q}^2\bar{p}^2 + \mathcal{O}(\bar{q}^6 + \bar{p}^6).$$

This provides us with a correct expansion up to order 5 of a reduced canonical Hamiltonian on the center manifold.

4.2 Reduction of Poisson structures

We now consider the case J not being invertible. We then call (4.1) a generalized Hamiltonian system. Our first result does not even assume $J^* = -J$; hence it applies also to gradient systems, where J is the identity I_X (see Section 6.1), or to the so-called dissipative Hamiltonian systems discussed in [Gr86].

We make the following assumptions on $J = \left(\begin{smallmatrix} J_1 & J_{12} \\ J_{21} & J_2 \end{smallmatrix}\right)$:

(A1): J is of class C^2 from X into $\mathcal{L}(X^*, X)$.

(A2): $J_2(0) : X_2^* \to X_2$ has a bounded inverse.

Remark: The theory developed below could be generalized to cases with unbounded operators J, which relates to weakly nondegenerate symplectic forms.

Theorem 4.4

Assume that **(A1)** and **(A2)** hold and that the system (4.1) has a locally invariant C^2–submanifold of the form $x_2 = h(x_1)$ with $h(0) = Dh(0) = 0$. Then the reduced equation (4.3) has the form

$$\dot{x}_1 = \tilde{J}(x_1)\mathbf{d}\tilde{H}(x_1), \qquad (4.5)$$

where $\tilde{H}(x_1) = H(x_1 + h(x_1))$ and

$$\tilde{J}(x_1) = J_1 + (-J_1 B^* + J_{12})(J_2 - BJ_{12} - J_{21}B^* + BJ_1 B^*)^{-1}(BJ_1 - J_{21})$$

with $B = Dh(x_1)$ and $J_i = J_i(x_1 + h(x_1))$.

Proof: From the invariance of M_C and from (4.1) we obtain

$$J_{21}\mathbf{d}_1 H + J_2 \mathbf{d}_2 H = \dot{x}_2 = B\dot{x}_1 = B(J_1 \mathbf{d}_1 H + J_{12}\mathbf{d}_2 H)$$

where $\mathbf{d}_1 H(v_1) + \mathbf{d}_2 H(v_2) = \mathbf{d}H(v_1 + v_2)$. Using $\mathbf{d}\tilde{H} = \mathbf{d}_1 H + B^* \mathbf{d}_2 H$ yields

$$\mathbf{d}_2 H = \hat{J}(BJ_1 - J_{21})\mathbf{d}\tilde{H}, \quad \mathbf{d}_1 H = [I - B^*\hat{J}(BJ_1 - J_{21})]\mathbf{d}\tilde{H},$$

where $\hat{J} = (J_2 - BJ_{12} - J_{21}B^* + BJ_1 B^*)^{-1}$. Note that $\hat{J}$ exists in a neighborhood of $x = 0$ by **(A2)** and the relation $B(x_1) = \mathcal{O}(\|x_1\|)$. Inserting this into (4.1) gives exactly the equation (4.5).   □

To see the relations to the symplectic case we compare the formula in Theorem 4.4 with equation (4.4) in the special case $J = \begin{pmatrix} J_1 & 0 \\ 0 & J_2 \end{pmatrix} = \begin{pmatrix} \Omega_1^{-1} & 0 \\ 0 & \Omega_2^{-1} \end{pmatrix}$. Then we obtain

$$\begin{aligned}
\tilde{J} &= J_1 - B^*(J_2 + BJ_1 B^*)^{-1}BJ_1 = J_1 - J_1 B^*(J_2(I + \Omega_2 BJ_1 B^*))^{-1}BJ_1 \\
&= J_1 - J_1 B^* \sum_{k=0}^{\infty}(-\Omega_2 BJ_1 B^*)^k \Omega_2 BJ_1 = J_1 - J_1 \sum_{k=0}^{\infty}(-1)^k(B^*\Omega_2 BJ_1)^{k+1} \\
&= J_1(I + B^*\Omega_2 BJ_1)^{-1} = ((I + B^*\Omega_2 BJ_1)\Omega_1)^{-1} = \tilde{\Omega}^{-1}.
\end{aligned}$$

The question remains whether $\tilde{J}$ defines a Poisson structure, if J does so. First we give a Lemma which relates the Jacobi identity in (2.5) to an equation involving J directly.

Lemma 4.5

A function $J : X \to \mathcal{L}(X^*, X)$ defines a Poisson bracket $\{\,\cdot\,,\,\cdot\,\}$ if and only if

$$\langle a_1, J(x)\mathbf{d}\langle a_2, J(x)a_3\rangle\rangle + \langle a_2, J(x)\mathbf{d}\langle a_3, J(x)a_1\rangle\rangle + \langle a_3, J(x)\mathbf{d}\langle a_1, J(x)a_2\rangle\rangle = 0 \quad (4.6)$$

for all $x \in X$ and all $a_i \in X^*$ and $J^* = -J$.

Proof: To show that (4.6) is a consequence of $(2.5)_4$ we simply take $F_i(x) = \langle a_i, x \rangle$. On the other hand, if we take a fixed $x = x_0$ and let $a_i = \mathbf{d}F_i(x_0)$, then $(2.5)_4$ has the form

$$0 = \langle a_1, J(x)\mathbf{d}\langle a_2, J(x)a_3 \rangle \rangle + \langle a_1, J(x_0)[D^2 F_3 J^* a_2 + (D^2 F_2)^* J a_3] \rangle + \text{ cyclic perm.}$$

Here the sum of the first terms, involving a_i only, cancels due to (4.6), while the terms with the second deviatives in F cancel since $D^2 F_i = (D^2 F_i)^*$ and $J = -J^*$. $\qquad\square$

Theorem 4.6
Assume that J satisfies **(A1)** *and* **(A2)** *and defines a Poisson bracket on X. Then, $\widetilde{J}$ as given in (4.5) defines a reduced Poisson structure on a neighborhood $U_1 \subset X_1$ of $x_1 = 0$.*

Proof: The skew–symmetry of J implies $J_1^* = -J_1$, $J_2^* = -J_2$, and $J_{21} = -J_{12}^*$. Hence the skew–symmetry of $\widetilde{J}$ follows immediately.

To proof the relation (4.6), we first transform it into a more convenient form. As $a_i \in X^*$ are independent of $x \in X$, we have

$$\langle a_1, J(x)\mathbf{d}\langle a_2, J(x)a_3 \rangle \rangle = -\langle J(x)a_1, \mathbf{d}\langle a_2, J(x)a_3 \rangle \rangle =$$
$$-\mathbf{d}\langle a_2, J(x)a_3 \rangle (J(x)a_1) = -\langle a_2, DJ[Ja_1]a_3 \rangle$$

where $DJ[v]$ denotes the directional derivative $\lim_{t\to 0} \frac{1}{t}[J(x+tv) - J(x)]$. Using the short notation $J_v = DJ[v]$ the condition (4.6) is now equivalent to

$$\langle a_1, J_{Ja_2}a_3 \rangle + \langle a_2, J_{Ja_3}a_1 \rangle + \langle a_3, J_{Ja_1}a_2 \rangle = 0, \qquad (4.7)$$

for all $x \in X$ and all $a_i \in X^*$.

Of course the proof that $\widetilde{J}$ satisfies Jacobi's identity (4.7) has to be transfered back to Jacobi's identity of J. Therefore we define for $a \in X_1^*$ the vector $\widetilde{a} = \widetilde{J}a + B\widetilde{J}a \in TM_C$. Since all the terms J_{ik} in the definition of $\widetilde{J}$ are evaluated on M_C (i.e. they depend on $x_1 + h(x_1)$), the derivative $\widetilde{J}_{\widetilde{J}a}$ will involve the terms $J_{ik\widetilde{a}}$ and $B_{\widetilde{J}a}$ only. In particular we have

$$\begin{aligned}
\widetilde{J}_{\widetilde{J}a} &= I(a) + K(a), \\
I(a) &= C^*\widehat{J}B_{\widetilde{J}a}J_1 - C^*\widehat{J}(J_{12}^* B_{\widetilde{J}a}^* + B_{\widetilde{J}a}J_1 B^* - B_{\widetilde{J}a}J_{12} + BJ_1 B_{\widetilde{J}a}^*)\widehat{J}C - J_1 B_{\widetilde{J}a}^*\widehat{J}C \\
&= C^*\widehat{J}B_{\widetilde{J}a}\widetilde{J} - \widetilde{J}B_{\widetilde{J}a}^*\widehat{J}C, \\
K(a) &= J_{1\widetilde{a}} + (-J_{1\widetilde{a}}B^* + J_{12\widetilde{a}})\widehat{J}C - C^*\widehat{J}(J_{2\widetilde{a}} - BJ_{12\widetilde{a}} + J_{12\widetilde{a}}^* B^* + BJ_{1\widetilde{a}}B^*)\widehat{J}C \\
&\quad + C^*\widehat{J}(BJ_{1\widetilde{a}} + J_{12\widetilde{a}}^*).
\end{aligned}$$

We have sorted the terms such that I contains the derivatives of B and K those of J. Using the symmetry of $D^2 h$, $B_v w = B_w v$, yields

$$\langle a_2, I(a_1)a_3 \rangle = \langle a_2, C^*\widehat{J}B_{\widetilde{J}a_1}\widetilde{J}a_3 \rangle - \langle a_2, \widetilde{J}B_{\widetilde{J}a_1}^*\widehat{J}Ca_3 \rangle$$
$$= -\langle \widehat{J}Ca_2, B_{\widetilde{J}a_1}\widetilde{J}a_3 \rangle + \langle B_{\widetilde{J}a_1}\widetilde{J}a_2, \widehat{J}Ca_3 \rangle = -\langle \widehat{J}Ca_2, B_{\widetilde{J}a_3}\widetilde{J}a_1 \rangle + \langle \widehat{J}Ca_3, B_{\widetilde{J}a_1}\widetilde{J}a_2 \rangle.$$

It follows immediately that $\langle a_2, I(a_1)a_3 \rangle + $ cyclic perm. $= 0$. On the other hand

$$
\begin{aligned}
\langle a_2, K(a_1)a_3 \rangle_{X_1} &= \langle (I - B^* \widehat{J} C)a_2, J_{1\widetilde{a}_1}(I - B^* \widehat{J} C)a_3 \rangle + \langle (I - B^* \widehat{J} C)a_2, J_{12\widetilde{a}_1} \widehat{J} C a_3 \rangle \\
&\quad + \langle \widehat{J} C a_2, -J^*_{12\widetilde{a}_1}(I - B^* \widehat{J} C)a_3 \rangle + \langle \widehat{J} C a_2, J_{2\widetilde{a}_1} \widehat{J} C a_3 \rangle \\
&= \langle b_2, J_{\widetilde{a}_1} b_3 \rangle_X,
\end{aligned}
$$

where $b_i = (I - B^* \widehat{J} C)a_i + \widehat{J} C a_i \in X^*$. Moreover, the identities $\widetilde{J} = J_1(I - B^* \widehat{J} C) + J_{12} \widehat{J} C$ and $B\widetilde{J} = -J^*_{12}(I - B^* \widehat{J} C) + J_2 \widehat{J} C$ reveal that $\widetilde{a}_i = \widetilde{J} a_1 + B\widetilde{J} a_i = J b_i$. Thus, Jacobi's identity for $J = J(x)$ in the form of (4.7) results in

$$
\langle a_2, K(a_1)a_3 \rangle + \text{ cyclic perm. } = \langle b_2, J_{Jb_1} b_3 \rangle + \text{ cyclic perm. } = 0,
$$

which proves the assertion. $\square$

Remark 4.7 Note that the reduction theorems 4.1 and 4.6 do not require the existence of the equilibrium $x = 0$. The only important points are the invariance of the submanifold and the proper splitting of the Poisson structure stated in assumption **(A2)**. We will use this fact in Section 6.7.

Condition **(A2)** can be formulated coordinate independent similarly to Eqn. (2.2) in [MR86]. Given any submanifold $\mathcal{M}$ of $\mathcal{X}$ and a point $x \in \mathcal{M}$ we define the annihilator $E_x \mathcal{M} \subset T^*_x \mathcal{X}$ through

$$
E_x \mathcal{M} = \{ \alpha \in T^*_x \mathcal{X} \; : \; \langle \alpha, v \rangle = 0 \text{ for all } v \in T_x \mathcal{M} \}.
$$

Hence, $E\mathcal{M}$ defines a bundle over $\mathcal{M}$. Now $J(x)E_x \mathcal{M}$ is a subspace of $T_x \mathcal{X}$; and **(A2)** is equivalent to

$$
J(x)E_x \mathcal{M} + T_x \mathcal{M} = T_x \mathcal{X} \tag{4.8}
$$

for $x = 0$. This is seen most easily by using the splitting $T_0 \mathcal{X} = X_1 \times X_2$ with $X_1 = T_0 \mathcal{M}$. Thus $v \in T_0 \mathcal{X}$ has the form $v = (v_1, 0)$, and $\alpha \in E_0 \mathcal{M}$ implies $\alpha = (0, \alpha_2)$. Now $T_0 \mathcal{M} + J(0)E_0 \mathcal{M}$ contains elements $(v_1, 0) + (J_{12}(0)\alpha_2, J_2(0)\alpha_2)$ and the desired equivalence is proved.

Obviously, the condition (4.8) is independent of the splitting and can be formulated at each point $x \in \mathcal{M}$. If it is satisfied on all of $\mathcal{M}$, then we obtain a reduced Poisson structure on $\mathcal{M}$, by doing the local construction given above in appropriate local coordinate systems.

4.3   Flattening of center manifolds

We now give another way of effectively calculating the reduced system on the center manifold in a canonical Hamiltonian system. The general idea is to introduce new canonical coordinates in the full system such that the center manifold is flat in these coordinates. Then the reduced symplectic structure will be automatically constant on the center manifold. Thus, the use of Darboux's theorem can be avoided. For calculating the reduced problem up to a given order, this method seems to need less algebraic manipulations as the one presented in the first section, moreover it is straight forward and no arbitraryness appears. Thus, it is easy to write a computer algorithm to do the reduction automatically. This was already done in [Si90] for a special problem in celestial mechanics. Another important application of the flattening procedure appears in the reduction theory of Langrangian systems in Chapter 6. There it will prove the validity of the *relaxed natural reduction method*.

In Section 5.4 we establish the following general result:

Theorem 5.14
Let $\mathcal{X}$ be a symplectic manifold modelled over a Hilbert space and let $\mathcal{M}$ be a symplectic submanifold. Then, around each $m \in \mathcal{M}$ there exist local canonical coordinates (q_1, q_2, p_1, p_2) such that $\mathcal{M}$ is given by $(q_2, p_2) \equiv 0$.

This result applies in particular to a center manifold $\mathcal{M}_C$. Let $H = H(q_1, q_2, p_1, p_2)$ be the Hamiltonian in these coordinates. The invariance of $\mathcal{M}_C$ implies that X_H is tangential to $\mathcal{M}_C$, viz.

$$\mathbf{d}_{q_2} H(q_1, 0, p_1, 0) = 0, \ \mathbf{d}_{p_2} H(q_1, 0, p_1, 0) = 0 \quad \text{for all } q_1, p_1. \tag{4.9}$$

The converse is also true, i.e. if we find canonical coordinates such that (4.9) holds, then $(q_2, p_2) \equiv 0$ is an invariant submanifold.

Note that the reduction onto the center manifold is trivial in these coordinates: The reduced Hamiltonian is $\widetilde{H}(q_1, p_1) = H(q_1, 0, p_1, 0)$ and the center manifold coordinates (q_1, p_1) are already canonical.

However, for the construction of the coordinates (q_1, q_2, p_1, p_2) it is essential to know the center manifold $\mathcal{M}_C$ in the original coordinates explicitly. We now describe a method which relies on the same idea of flattening, but does not need the knowledge of the center manifold. In fact, it produces approximations of $\mathcal{M}_C$. Like in nonlinear normal form theory (cf. e.g. [Br88]) we successively remove all terms in the derivatives of (4.9), up to a given order by canonical transformations.

To avoid technicalities we restrict ourselves, in the remainder of this section, finite dimensional canonical systems with $\mathcal{X} = \mathbb{R}^{2l}$. According to the results of the previous chapter we may assume that canonical coordinates are given as $(x_1, x_2) = (q, p, Q, P)$ where (q, p) span the center space and (Q, P) the hyperbolic space. The Hamiltonian has the expansion

$$H(x_1, x_2) = H_2 + R = \frac{1}{2}\langle x_1, A_1 x_1\rangle + \frac{1}{2}\langle x_2, A_2 x_2\rangle + R(x_1, x_2); \quad A_j = A_j^*, \ R = \mathcal{O}(|x|^3).$$

As above the center manifold is given in the form $x_2 = h(x_1)$ and our aim is to make h of high order, then Theorem 4.4 shows that the reduced symplectic structure is very close to the canonical one (see Theorem 4.2). To this end we notice that $\dot{x}_2 = J_2(A_2 x_2 + \mathbf{d}_2 R(x_1, x_2))$. If we could transform coordinates such that $\mathbf{d}_2 R(x_1, 0) = 0$ for all x_1 we would have immediately that $x_2 \equiv 0$ is an invariant manifold and hence a center manifold. Here, we only show that it is possible to make the center manifold as flat as one like, i.e. $h(x_1) = \mathcal{O}(|x_1|^n)$ for arbitrary n. Since h satisfies the differential equation

$$Dh(x_1)[K_1 x_1 + f_1(x_1, h(x_1))] = K_2 h(x_1) + f_2(x_1, h(x_1))$$

we see that $f_2(x_1, x_2) = \mathcal{O}(|x_1|^n + |x_1|\,|x_2| + |x_2|^2)$ implies $h = \mathcal{O}(|x_1|^n)$. Thus, to flatten the center manifold we have to transform coordinates such that R does not contain terms being linear in $x_2 = (Q, P)$ up to the desired order of approximation, i.e. $\mathbf{d}_2 R(x_1, 0) = \mathcal{O}(|x_1|^n)$.

To reach any given n we construct the canonical transformations by induction over $m = 2, \ldots, n-1$; the case $m = 1$ is the trivial starting point. We assume that R has the form

$$R(x_1, x_2) = R(x_1, 0) + \langle x_2, \rho_m(x_1)\rangle + \mathcal{O}(|x_2|^2 + |x_1|^{m+1}|x_2|),$$

where ρ_m is homogeneous of degree $m \geq 2$. We consider canonical changes of coordinates $(q, p, Q, P) = T(\widehat{q}, \widehat{p}, \widehat{Q}, \widehat{P})$ derived from a *generating function* $S = S(q, \widehat{p}, Q, \widehat{P})$ by solving the relations $(p, \widehat{q}, P, \widehat{Q}) = \mathbf{d}S(q, \widehat{p}, Q, \widehat{P})$, see [Ar78] for a general treatment of generating functions. We look for S in the form

$$S(q, \widehat{p}, Q, \widehat{P}) = q \cdot \widehat{p} + Q \cdot \widehat{P} + S_m(q, \widehat{p}, Q, \widehat{P}) \text{ with } S_m = \langle (Q, \widehat{P}), \sigma_m(q, \widehat{p})\rangle,$$

where σ_m again is homogeneous of degree m. Inserting the induced transformation we obtain the new Hamiltonian $\widehat{H} = H_2 + \widehat{R}$ with

$$\widehat{R}(\widehat{x}_1, \widehat{x}_2) = R(\widehat{x}_1, \widehat{x}_2) + \{H_2, S_m\} + \mathcal{O}(|\widehat{x}|^{m+2}).$$

Using the form of S_m and comparing the coefficients linear in x_2 we find that $\mathbf{d}_2 \widehat{R}(x_1, 0)$ is of order $\mathcal{O}(|x_1|^{m+1})$ if and only if σ_m satisfies

$$A_2 J_2 \sigma_m(x_1) + D_{x_1}\sigma_m(x_1)[J_1 A_1 x_1] = -\rho_m(x_1).$$

This system is always solvable and the solution can be given as follows:

$$\sigma_m(x_1) = \mathcal{G}(\sigma_m)(x_1) = -\int_{t\in\mathbf{R}} \widehat{G}_2(t)\rho_m(e^{J_1 A_1 t}x_1)\,dt. \tag{4.10}$$

Here $\widehat{G}_2$ is the Green's function of the linear problem $\dot{x}_2 = A_2 J_2 x_2 + f_2(t)$ on the space of bounded function. The proof is the same as for (2.11), and $\widehat{G}_2(t) = -G_2^*(-t)$ since $K_2^* = (J_2 A_2)^* = -A_2 J_2$. Note that even in the infinite dimensional case, where H is only defined on a manifold domain, each step can be carried through as in case of constructing the Taylor expansion of the reduction function.

Thus, we are able to remove the terms ρ_m successively for each $m = 2, \ldots, n-1$ and obtain a Hamiltonian $\widehat{H}(\widehat{x}_1, \widehat{x}_2) = \widehat{H}(\widehat{x}_1, 0) + \mathcal{O}(|\widehat{x}_1|^n |\widehat{x}_2| + |\widehat{x}_2|^2)$ and a center manifold satisfying $\widehat{x}_2 = h(\widehat{x}_1) = \mathcal{O}(|\widehat{x}_1|^n)$. The reduced Hamiltonian has the expansion $\widetilde{H}(\widehat{x}_1) = \widehat{H}(\widehat{x}_1, 0) + \mathcal{O}(|\widehat{x}_1|^{2n})$. Moreover, as in Theorem 4.2 we obtain $\widetilde{\Omega}(\widehat{x}_1) = \begin{pmatrix} 0 & I \\ -I & 0 \end{pmatrix} + \mathcal{O}(|\widehat{x}_1|^{2n-2})$ for the reduced symplectic structure. According to Theorem 5.15 we then find canonical coordinates $(\overline{q}, \overline{p}) = (\widehat{q}, \widehat{p}) + \mathcal{O}(|(\widehat{q}, \widehat{p})|^{2n-1})$. The associated Hamiltonian $\overline{H}$ reads $\overline{H}(\overline{q}, \overline{p}) = \widehat{H}(\overline{q}, \overline{p}, 0, 0) + \mathcal{O}(|(\overline{q}, \overline{p})|^{2n})$.

Summarizing we have the following result.

Theorem 4.8
Let $H = H(x) \in C^k(\mathbb{R}^{2l}, \mathbb{R})$ be a Hamiltonian of a canonical Hamiltonian systems having an equilibrium in $x = 0$ with a (non-trivial) center manifold. Then for any $n < k$ there exists an analytical canonical transformation $x = T(q, Q, p, P)$ such that the center manifold is given by $(Q, P) = h(q, p) = \mathcal{O}(|(q, p)|^n)$. Taking (q, p) as canonical coordinates on the center manifold and using $\overline{H}(q, p) = H(T(q, 0, p, 0))$ as Hamiltonian gives the correct terms up to order $2n$ of the true reduced Hamiltonian system.

Example: To illustrate the method we return to the example of Section 4.1:
$$H(q, Q, p, P) = \tfrac{1}{2}(p^2 + \nu^2 q^2) + QP + q^2(\alpha Q + \beta P).$$

and give the calculations for the expansion of the system by the method just described.

We want to transform the Hamiltonian by a coordinate change such that the terms being just linear in (Q, P) are of high order in (q, p). Hence, we make an ansatz for the generating function S in the form
$$S(q, \widehat{p}, Q, \widehat{P}) = q\widehat{p} + Q\widehat{P} + Q(s_1 q^2 + s_2 q\widehat{p} + s_3 \widehat{p}^2) + \widehat{P}(s_4 q^2 + s_5 q\widehat{p} + s_6 \widehat{p}^2).$$

The equation for σ_2 now reduces to
$$\begin{pmatrix} -1 & 0 \\ 0 & 1 \end{pmatrix} \begin{pmatrix} s_1 q^2 + s_2 qp + s_3 p^2 \\ s_4 q^2 + s_5 qp + s_6 p^2 \end{pmatrix} + \begin{pmatrix} 2s_1 q + s_2 p & s_2 q + 2s_3 p \\ 2s_4 q + s_5 p & s_5 q + 2s_6 p \end{pmatrix} \begin{pmatrix} p \\ -\nu^2 q \end{pmatrix} = -\begin{pmatrix} \alpha q^2 \\ \beta q^2 \end{pmatrix}.$$

We obtain $(s_1, s_2, s_3) = \beta \nu_4^{-1}(\nu_2, 2, 2)$ and $(s_4, s_5, s_6) = \alpha \nu_4^{-1}(-\nu_2, 2, -2)$ (recall $\nu_i = 1 + i\nu^2$). The induced transformation is
$$\begin{pmatrix} q \\ p \\ Q \\ P \end{pmatrix} = \begin{pmatrix} \widehat{q} \\ \widehat{p} \\ \widehat{Q} \\ \widehat{P} \end{pmatrix} + \begin{pmatrix} -(s_2\widehat{q} + 2s_3\widehat{p})\widehat{Q} - (s_5\widehat{q} + 2s_6\widehat{p})\widehat{P} \\ (2s_1\widehat{q} + s_2\widehat{p})\widehat{Q} + (2s_4\widehat{q} + s_5\widehat{p})\widehat{P} \\ -(s_4\widehat{q}^2 + s_5\widehat{q}\widehat{p} + s_6\widehat{p}^2) \\ s_1\widehat{q}^2 + s_2\widehat{q}\widehat{p} + s_3\widehat{p}^2 \end{pmatrix} + \mathcal{O}(|(\widehat{q}, \widehat{p})|^2 |(\widehat{Q}, \widehat{P})|)$$

and the transformed Hamiltonian has the expansion

$$
\begin{aligned}
\widehat{H}(\widehat{q},\widehat{p},\widehat{Q},\widehat{P}) \;=\; & \tfrac{1}{2}(\widehat{p}^2 + \nu^2\widehat{q}^2) + \widehat{Q}\widehat{P} + \nu_2\nu_4^{-1}(\alpha^2 + \beta^2 + \nu_2\nu_4^{-1}\alpha\beta)\widehat{q}^4 + \\
& +2\nu_4^{-1}(\beta^2 - \alpha^2)\widehat{q}^3\widehat{p} + 2\nu_4^{-1}(\alpha^2 + \beta^2 + 2\nu_2\nu_4^{-1}\alpha\beta)\widehat{q}^2\widehat{p}^2 + \\
& +0\widehat{q}^3\widehat{p} + 4\nu_4^{-2}\alpha\beta\widehat{p}^4 + \mathcal{O}(|(\widehat{q},\widehat{p})|^3|(\widehat{Q},\widehat{P})| + |(\widehat{Q},\widehat{P})|^2).
\end{aligned}
$$

Thus, in these coordinates the center manifold satisfies the relation $(\widehat{Q},\widehat{P}) = h(\widehat{q},\widehat{p}) = \mathcal{O}(|(\widehat{q},\widehat{p})|^3)$ and we obtain the same order of approximation of the reduced problem as in Section 4.1: $\overline{H}(\overline{q},\overline{p}) = \widehat{H}(\overline{q},\overline{p},0,0) + \mathcal{O}(|(\overline{q},\overline{p})|^6)$.

Since in the elimination procedure developed above no small divisors appear one could hope that for analytical systems the flattening can be carried through to any order and that in the limit $n \to \infty$ the system and the associated canonical transformation converge to a limit. Note that this procedure does not transform the system into normal form. For the normal form transformation it is known that it does not converge in general, see [Br89]. If the flattening procedure would converge, we would have found a way to show that the center manifold is *analytical* for analytical Hamiltonian systems. In general this is not true, as is seen in the following example with four degrees of freedom. This question of finding analytical invariant submanifolds is called **Problem 1** in [Br89, pg.202].

We let $(q,p) = (q_1, q_2, \varepsilon, \delta, p_1, p_2, p_\varepsilon, p_\delta)$ and

$$
H(q,p) = \frac{1}{2}(p_1^2 - \varepsilon^2 q_1^2 + p_2^2 - q_2^2) + \frac{1}{4}q_1^4 + \delta q_1^2 q_2.
$$

Note that p_ε and p_δ do not appear, and hence ε and δ may be considered as parameters in the system.

The origin is a fixed point with a linearization having the eigenvalues 1, -1, and 0, the last being 6-fold. The center manifold of the system is given by

$$
q_2 = h_1(q_1, p_1, \varepsilon, \delta), \quad p_2 = h_p(q_1, p_1, \varepsilon, \delta)
$$

where $(q_1, p_1, \varepsilon, \delta) \in U \subset \mathbb{R}^4$. It is easy to see that h does not depend on $(p_\varepsilon, p_\delta)$. If we now assume that the center manifold for this problem is even analytic in a neighborhood of zero then we have an analytic center manifold for the four–dimensional parameter-dependent system

$$
\begin{aligned}
\ddot{q}_1 - \varepsilon^2 q_1 + q_1^3 + 2\delta q_1 q_2 &= 0 \\
\ddot{q}_2 - q_2 + \delta q_1^2 &= 0,
\end{aligned}
\tag{4.11}
$$

which even depends analytically on (ε, δ).

We will show that (4.11) cannot have an analytical center manifold for $\varepsilon = 1/m$, $m \in \mathbb{N}$, and δ small. For $\delta = 0$ the center manifold is given by $(q_2, p_2) \equiv 0$, and on the center manifold a pair of homoclinic orbits exists: $q_1(t) = \pm\mu(t, \varepsilon) = \pm\varepsilon\sqrt{2}/\cosh(\varepsilon t)$. These orbits persist for sufficiently small δ and ε fixed; they still lie on the center manifold, since they are small bounded solutions. Let us call this family of solutions $\tilde{q}(t, \varepsilon, \delta)$, then from (4.11) we obtain

$$
\begin{aligned}
\tilde{q}_1(t, \varepsilon, \delta) &= \mu(t, \varepsilon) + \mathcal{O}(\delta^2), \\
\tilde{q}_2(t, \varepsilon, \delta) &= \delta \int_{\mathbf{R}} \frac{1}{2} e^{-|t-s|} \mu(s, \varepsilon) \, ds + \mathcal{O}(\delta^3).
\end{aligned}
$$

Note that for $t \geq 1$ the function μ has the absolutely convergent expansion $\mu(t, \varepsilon) = 2\varepsilon\sqrt{2} \sum_0^\infty (-1)^n e^{-(2n+1)\varepsilon t}$. Hence, in the case $\varepsilon = 1/m$ the integral in the expansion for q_2 has, for $t \geq 1$, the form

$$
\frac{\partial}{\partial \delta} \tilde{q}_2(t, \varepsilon, 0) = \sum_{n \neq m} (-1)^n \frac{2}{1 - (n/m)^2} e^{-(2n+1)t/m} + e^{-t}(r_m + (-1)^m(t - 1/2)),
$$

where $r_m = \int_{-\infty}^1 e^s \mu \, ds$. The appearance of the term te^{-t} will enable us to prove non–analyticity as follows. If the center manifold is analytic then the homoclinic solution $\tilde{q}_1$ solves an analytic system on the center manifold. Since the eigenvalues of the reduced linear system are exactly $\pm\varepsilon$, independently of δ, the homoclinic solution has a convergent expansion in powers of $e^{-n\varepsilon t}$ for sufficiently large t. From the analyticity of $q_2 = h_1(q_1, p_1, \varepsilon, \delta)$ the same must hold for $\tilde{q}_2$, however this contradicts the above expansion. Hence, the center manifold is not analytic. Note that all solutions on the center manifold are bounded in $(q, p) \in \mathbb{R}^4$ and form a two–dimensional manifold for each fixed ε and δ. Since any center manifold must contain all small bounded solutions, it has to be unique.

Nevertheless, there is some hope that center manifolds for analytic Hamiltonian systems are again analytic. If the center manifold is two–dimensional and is filled with periodic solutions of bounded period the the results is true, see [Mo58, Au85]. Moreover, by the results in [HS85] it is known that each small bounded solution on the center manifold depends analytically on the time variable.

We end this section with the following **conjecture:**

> If the center manifold of an analytic Hamiltonian system is completely filled
> with bounded solutions (or more restrictive: the reduced Hamiltonian has
> a positive (negative) definite second derivative at the fixed point), then the
> center manifold is analytic.

Clearly the definiteness of the second derivative implies boundedness of all solutions on the center manifold, and hence the uniqueness of the center manifold. Note that the above example is not a counter example, since only the reduced problem (4.11) has bounded solutions on the center manifold. The original Hamiltonian system including the variables p_ε and p_δ on the center manifold has unbounded solutions due to

$$\dot{p}_\varepsilon = -\frac{\partial}{\partial\varepsilon}H = -\varepsilon q_1^2, \quad \dot{p}_\delta = -\frac{\partial}{\partial\delta}H = -q_1^2 q_2.$$

For instance, we have the unbounded solution $(q,p) = (\varepsilon, 0, \varepsilon, 0, 0, 0, -\varepsilon^3 t, 0)$. Recall that in this case the center manifold is unique.

The conjecture is based on the following observation. For a convergence proof of the flattening transformations it is essential to control the growth of the auxiliary function σ_m in the generating function S of m-th order. To estimate this we have to know the norm of the solution operator $\mathcal{G}$ defined in (4.10). The linear theory gives $\|\widehat{G}_2(t)\| \leq C_2 e^{-\alpha|t|}$ and $\|e^{J_1 A_1 t}\| \leq C_1(1+|t|)^r$, where r is the maximum length of all Jordan chains corresponding to $K_1 = J_1 A_1$. On the space of homogeneous function we introduce the norm $\|\sigma_m\| = \max\{|\sigma_m(x)| : |x| \leq 1\}$, then the estimate

$$\begin{aligned}\|\mathcal{G}(\sigma_m)\| &\leq \int_{\mathbf{R}} C_2 e^{-\alpha|t|}\left(C_1(1+|t|)^r\right)^m \|\rho_m\|\, dt \\ &\leq \frac{2e^\alpha C_2}{\alpha}\left(\frac{C_1}{\alpha^r}\right)^m (rm)! \|\rho_m\|\end{aligned}$$

holds. We see that the operator norm grows faster than exponentially in m when $r \neq 0$. However, when the reduced Hamiltonian has a definite second derivative at the fixed point, then all eigenvalues are simple which implies $r = 0$. For this case it seems more reasonable to expect convergence, since the operator norm grows only exponentially.

Chapter 5

Hamiltonian systems with symmetries

In the previous chapter we have seen that reducing a canonical Hamiltonian system onto a center manifold, in general, leads to a non–canonical symplectic structure. If for the further analysis a canonical structure is needed, as for instance in the Lagrange formalism (see Chapter 6), the Darboux theorem can be used to find coordinates in which the symplectic form is canonical. Here we recall the basic aspects and generalize the theory to problems being invariant under the action of a Lie group.

5.1 Lie groups

First we give some background on the theory of Lie groups, for further details see [AM78, Ch.4]. A *Lie group* is a group which also has a manifold structure such that the group operations of multiplication $\cdot : G \times G \to G; (g, h) \to gh$ and the inversion $\cdot^{-1} : G \to G; g \to g^{-1}$ are smooth mappings. The identity is denoted by e. We only consider finite–dimensional Lie groups. The *left (right) translation* L_g (R_g): $G \to G$ is given by $h \to gh$ ($h \to hg$).

The tangent and cotangent spaces $T_e G$ and $T_e^* G$ at e are denoted by $\mathbf{g}$ and $\mathbf{g}^*$, respectively. $\mathbf{g}$ is called the *Lie algebra of G*. Let $\xi_1, \xi_2 \in \mathbf{g}$ and define the left–invariant vector fields v_1, v_2 on TG by the relation $v_{i\,g} = DL_g \xi_i \in T_g G$. Using the Lie bracket for vector fields on TG, as defined in Chapter 2, we obtain the vector field $w = [v_1, v_2]$ on TG. Now the *Lie bracket* $[\cdot, \cdot]_\mathbf{g}$ on the Lie algebra $\mathbf{g}$ is defined by

$$[\xi_1, \xi_2]_\mathbf{g} = [v_1, v_2]_{TG}(e) \in \mathbf{g} = T_e G.$$

It can be shown that $[\cdot, \cdot]_\mathbf{g} : \mathbf{g} \times \mathbf{g} \to \mathbf{g}$ satisfies the properties of a Lie bracket, viz. it is bilinear, skew–symmetric, and Jacobi's identity holds (cf. (2.5)). Subsequently we drop the index $\mathbf{g}$ at the Lie bracket as no confusion can occur.

The Lie bracket can also be defined by the use of the *exponential function* $\exp : \mathbf{g} \to G$. For $\xi \in \mathbf{g}$ we define $g(t) = \exp(t\xi)$, $t \in \mathbb{R}$, to be the solution curve of the differential equation $\dot{g} = DL_g\xi$, $g(0) = e$. Then the relation $\exp((t+s)\xi) = \exp(t\xi)\exp(s\xi)$ holds. Moreover we have

$$
\begin{aligned}
[\xi_1, \xi_2] &= \frac{d^2}{dt^2}\{\exp(t\xi_1)\exp(t\xi_2)\exp(-t\xi_1)\exp(-t\xi_2)\}_{t=0} \\
&= \frac{d^2}{ds\,dt}\{\exp(t\xi_1)\exp(s\xi_2)\exp(-t\xi_1)\exp(-s\xi_2)\}_{s,t=0}.
\end{aligned}
$$

For example, the set of invertible $n \times n$-matrices, $GL(n)$, forms under the usual matrix multiplication a Lie group. The associated Lie algebra is $L(\mathbb{R}^n, \mathbb{R}^n)$ with the Lie bracket $[A, B] = AB - BA$. Any subgroup of $GL(n)$, which is also a submanifold, provides new examples for Lie groups. For instance, the set

$$
\left\{ \begin{pmatrix} R & r \\ 0 & 1 \end{pmatrix} \in GL(4) \; : \; R \in SO(3), \; r \in \mathbb{R}^3 \right\}
$$

forms the six–dimensional Lie group of Euclidian transformations on $\mathbb{R}^3$. (A pair $(R, r) \in SO(3) \times \mathbb{R}^3$ maps $x \in \mathbb{R}^3$ into $y = Rx + r \in \mathbb{R}^3$.) The Lie algebra is given by

$$
\left\{ \begin{pmatrix} A & a \\ 0 & 0 \end{pmatrix} \in L(\mathbb{R}^4, \mathbb{R}^4) \; : \; A + A^* = 0, \; a \in \mathbb{R}^3 \right\}
$$

with the Lie bracket $[(A, a), (B, b)] = (AB - BA, Ab - Ba)$.

An *action* Φ of a Lie group G on the manifold $\mathcal{X}$ is a smooth mapping $\Phi : G \times \mathcal{X} \to \mathcal{X}$ such that $\Phi(g, \Phi(h, x)) = \Phi(gh, x)$ and $\Phi(e, x) = x$ for all $x \in \mathcal{X}$ and all $g, h \in G$. We also use the notation Φ_g for the mapping $\Phi(g, \cdot) : \mathcal{X} \to \mathcal{X}$ which is a diffeomorphism for each $g \in G$. If the manifold is a linear space X and if the mappings Φ_g are bounded linear operators the action is called a *linear representation of G on X*.

Note that $\varphi_g : h \to ghg^{-1}$ defines an action of G on itself. Taking the differential at $h = e$ we define the *adjoint action* Ad_g of G on $\mathbf{g}$ by $Ad_g\xi = D_h\varphi_g(e)[\xi]$. It can also be expressed by $Ad_g = DR_{g^{-1}}DL_g = DL_gDR_{g^{-1}}$. Moreover the *coadjoint action* $Ad^*_{g^{-1}}$ of G on $\mathbf{g}^*$ is given by $Ad^*_g = DL^*_gDR^*_{g^{-1}}$, or equivalently by $\langle Ad^*_{g^{-1}}\eta, \xi \rangle = \langle \eta, Ad_g\xi \rangle$, $\xi \in \mathbf{g}$ and $\eta \in \mathbf{g}^*$.

Furtheron we assume that the generalized Hamiltonian system

$$
\dot{x} = X_H(x) = J(x)\mathbf{d}H(x) \tag{5.1}
$$

is invariant under the action Φ of a Lie group G. By this we mean that H and the Poisson bracket $\{\cdot, \cdot\}$ (resp. ω) are separately invariant under the action Φ:

$$
H(\Phi_g(x)) = H(x), \qquad J(\Phi_g(x)) = D\Phi_g(x)J(x)D\Phi^*_g(x) \tag{5.2}
$$

for all $g \in G$ and all $x \in \mathcal{X}$. We shortly say that an object is *G–invariant*, if it is invariant under all transformations Φ_g. As a consequence, for every solution $x(t)$, $t \in (a, b)$, of (5.1) and every $g \in G$ the curve $\Phi_g(x(t))$, $t \in (a, b)$, is again a solution.

We now briefly describe the classical *reduction of Hamiltonian systems by symmetry*. For an exact treatment see [AM78, Ma81]. We do this to exhibit the different nature of this reduction compared to the center manifold reduction discussed below.

First we note that by Noether's theorem there is a momentum mapping $\mathcal{J} : \mathcal{X} \to \mathbf{g}^*$ such that $\mathcal{J}$ is constant along any solution of the Hamiltonian system. For a fixed $\mu \in \mathbf{g}^*$ we define the level set

$$\mathcal{J}^{-1}(\mu) = \{\, x \in \mathcal{X} \; : \; \mathcal{J}(x) = \mu \,\}$$

being invariant under the flow.

We further assume that $\mathcal{J}$ is Ad^*–equivariant, i.e. $\mathcal{J}(\Phi_g(x)) = Ad^*_{g^{-1}} \mathcal{J}(x)$. Let

$$G_\mu = \{\, g \in G \; : \; Ad^*_{g^{-1}} \mu = \mu \,\},$$

the isotropy group of $\mu \in \mathbf{g}^*$. Now the action Φ reduces to an action of G_μ on $\mathcal{J}^{-1}(\mu)$, and we are able to define the equivalence relation $\sim_\mu$ on $\mathcal{J}^{-1}(\mu)$ by

$$x \sim_\mu y \iff \exists g \in G_\mu : \Phi_g(x) = y.$$

The *reduced phase space* $\mathcal{X}_\mu$ is obtained by factoring $\mathcal{J}^{-1}(\mu)$ with respect to $\sim_\mu$. By G–invariance the Hamiltonian H has a well–defined restriction H_μ on $\mathcal{X}_\mu$ and, additionally, there is a unique induced symplectic structure ω_μ on $\mathcal{X}_\mu$.

For an Abelian group G the adjoint and the coadjoint action is trivial, hence we always have $G_\mu = G$. Then the reduced phase space $\mathcal{X}_\mu$ has codimension $2 \dim G$ in $\mathcal{X}$. This relates to the case of having a Hamiltonian $H = H(q_{m+1}, \ldots, q_n, p_1, \ldots, p_n)$ with m cyclic variables $(q_1, \ldots, q_m)$. Here the Lie group G is $I\!\!R^m$ with the action $\Phi_{(g_1, \ldots, g_m)}(q, p) = (q_1 + g_1, \ldots, q_m + g_m, q_{m+1}, \ldots, q_n, p)$. The Lie algebra and its dual can be identified with $I\!\!R^m$. For the momentum mapping we find $\mathcal{J}(q, p) = (p_1, \ldots, p_m)$. Thus, fixing $\mu \in \mathbf{g}^*$ means keeping $\mu = (p_1, \ldots, p_m)$ constant and factoring with respect to $\sim_\mu$ means just neglecting $(q_1, \ldots, q_m)$. Hence, in this case the reduction coincide with the standard reduction by cyclic variables.

The *center manifold reduction for symmetric system* will proceed in a completely different manner. First we will factor out the symmetry and construct a center manifold for the factored problem. By recovering the variables neglected in the factorization we then find a G–invariant symplectic center manifold with a G–invariant reduced Hamiltonian. Hence, the above described method of symmetry reduction can be applied to the reduced system on the center manifold. For an example see Saint–Venant's problem in Ch. 11.

5.2   Factorization by symmetry

From now on we will denote the full Lie group by $\widetilde{G}$ and G will only play the role of a specific subgroup. In the center manifold theory at an equilibrium $x_0 \in \mathcal{X}$ we have to distinguish two cases. Firstly there may exist $\widetilde{g} \in \widetilde{G}$ such that $\Phi_{\widetilde{g}}(x_0) = x_0$; and secondly we may have $\widetilde{g}$ with $\Phi_{\widetilde{g}}(x_0) \neq x_0$. Obviously the set

$$S = \{\, \widetilde{g} \in \widetilde{G} \; : \; \Phi_{\widetilde{g}}(x_0) = x_0 \,\}$$

is a subgroup of $\widetilde{G}$. All the other elements of $\widetilde{G}$ will move the point x_0 and we define

$$O_G(x) = \{\, \Phi_{\widetilde{g}}(x) \in \mathcal{X} \; : \; \widetilde{g} \in \widetilde{G}\,\},$$

the *orbit of* x under the action $\widetilde{G}$. To see the structure of the orbits we study the derivative $D_{\widetilde{g}}\Phi(e, x_0) : T_e\widetilde{G} \to T_{x_0}\mathcal{X}$, which vanishes when restricted to T_eS. However, there is always a subspace $N \subset T_e\widetilde{G}$ such that $\operatorname{Range} D_{\widetilde{g}}\Phi(e, x_0)|_N = \operatorname{Range} D_{\widetilde{g}}\Phi(e, x_0)$ and $N + T_eS = T_e\widetilde{G}$. In many cases N can be interpreted as the tangent space T_eG of a Lie subgroup G of $\widetilde{G}$. Here we make the following general assumptions on the action Φ of $\widetilde{G}$:

    a) $\widetilde{G}$ is the direct product of S and G ($\Phi_{sg} = \Phi_{gs}$).

    b) S is a compact Lie group with $\Phi_s(x_0) = x_0$ for all $s \in S$ (5.3)

    c) G is a Lie group such that $D_g\Phi(e, x_0) : T_eG \to T_{x_0}\mathcal{X}$ is injective.

For the direct product we shortly write $\widetilde{G} = G \odot S$. This means that for each $\widetilde{g} \in \widetilde{G}$ there are unique $s \in S$ and $g \in G$ such that $\widetilde{g} = sg$; moreover we have $sg = gs$ for all $s \in S$ and $g \in G$. This assumption allow us to decouple the transportational action of G from the action of S which rotates around x_0. This is a severe restriction but it simplifies the analysis considerably and it is always met in our applications.

Example: Let n and m be positive integers with $n \geq m$. Consider a bounded domain $\Sigma \subset I\!\!R^m$ such that $\int_\Sigma z_j dz = 0$ for all j. Denote by S the group of all the linear mappings $\tau : I\!\!R^m \to I\!\!R^m$ with $\tau(\Sigma) = \Sigma$. Hence, S is a subgroup of $O(m) = \{\, \tau \in I\!\!R^{m \times m} : \tau^*\tau = I \,\}$. In particular, S is a compact Lie group. If Σ is a ball in $I\!\!R^m$, then $S = O(m)$.

Now consider the Banach space $X = [L_p(\Sigma)]^n$, $p \in [1, \infty]$, with elements $u : \Sigma \to I\!\!R^n$ having components $(u_1, \ldots, u_n)$. On $I\!\!R^n$ consider the Lie group of all *Euclidian transformations*, $G = SO(n) \triangleright I\!\!R^n$. This means an element $(R, r) \in G$ defines the transformation $u \in I\!\!R^n \to Ru + r \in I\!\!R^n$. The group multiplication is the composition of these transformations, hence $(R_1, r_1) \cdot (R_2, r_2) = (R_1R_2, r_1 + R_1r_2)$.

Let $\widetilde{G}$ be the direct product $G \odot S$ and define the action $\Phi = \Phi_{(R,r,\tau)}$ of $\widetilde{G}$ on X by

$$\left(\Phi_{(R,r,\tau)}(u)\right)(z) = R\begin{pmatrix} \tau\overline{u}(\tau^*z) \\ \widehat{u}(\tau^*z) \end{pmatrix} + r, \qquad z \in \Sigma.$$

Here we have set $\bar{u} = (u_1, \ldots, u_m)$ and $\hat{u} = (u_{m+1}, \ldots, u_n)$.

Any function $u^0 \in X$ with $(\bar{u}^0(z), \hat{u}^0(z)) = (\alpha z, \beta)$ is mapped onto itself by the action of S. It remains to be checked whether condition (5.3c) holds. The Lie group of G is $\mathbf{g} = so(n) \triangleright {I\!\!R}^3 = \{(A, a) : A^* + A = 0\}$. Hence,

$$D_{(R,r)}\left(\Phi_{(R,r,I)}(u^0)\right)_{(R,r)=(I,0)} [(A, a)] = Au^0 + a.$$

For $\alpha \neq 0$ this spans a subspace of X of the $\dim(G) = n(n+1)/2$. Hence, for this example all the conditions (5.3) are satisfied. An application with $n = 3$ and $m = 2$ is the beam problem in Chapter 11.

Note that the symmetries with respect to G and S play a completely different role. The invariance under G allows us to factor the manifold $\mathcal{X}$ and the vector field X_H by G to obtain a reduced problem of codimension equal to $\dim(G)$. In doing this the symplectic structure will be reduced to a Poisson structure, and the reduced problem will still be invariant under the action of S. Denoting the linearization of the flow around x_0 by Kx_0 the symmetry under G induces an eigenvalue 0 of multiplicity $\dim G$, since $O_G(x_0)$ is a manifold of equilibria implying that $T_{x_0}O_G(x_0)$ is contained in the kernel of K. The symmetry under S induces the equivariance condition $D_x\Phi_s(x_0)K = KD_x\Phi_s(x_0)$ for all $s \in S$. Hence, one expects S to generate multiple eigenvalues: if $Kw = \lambda w$ then $D_x\Phi_s(x_0)w$ is also an eigenvector with eigenvalue λ, and if it is not a scalar multiple of w then λ has to be at least a double eigenvalue.

To establish the earlier mentioned decoupling of the actions of G and S we use a *slice theorem* as developed in [GS84, Ch.II.27], [Kr90] the references therein. However our version is different as we want to deal with noncompact Lie groups, at least for the subgroup G.

Theorem 5.1
(Slice theorem)
Let Φ be an action of the Lie group G on $\mathcal{X}$ such that the condition (5.3) holds. Then there is a linear closed subspace $Y \subset X$ with codimension $\dim G$ in X, there are neighborhoods $U \subset G$ of e, $V \subset \mathcal{X}$ of x_0, and $W \subset Y$ of 0, and there is a diffeomorphism $\theta : V \to U \times W$ such that the induced action $\Psi_{\tilde{g}} = \theta \circ \Phi_{\tilde{g}} \circ \theta^{-1}$ of $\tilde{G}$ on $G \times Y$ satisfies

$$\Psi_{sg_1}(g_2, y) = (g_1g_2, \tau_s y), \tag{5.4}$$

where τ_s, $s \in S$, is a linear representation of S on Y. (Note that the action on G is just the left translation of the G–component, and that the actions of G and S are decoupled.)

Proof: By a local coordinate system we may identify $\mathcal{X}$ and $T_0\mathcal{X}$ with X and x_0 with 0. We still denote the induced action of $\tilde{G}$ on X by Φ.

For $s \in S$ we have $\Phi_s(0) = 0$. The relation $\Phi_g(0) = \Phi_{gs}(0) = \Phi_s(\Phi_g(0))$ yields, after differentiation with respect to g at $g = e$, that $D_x\Phi_s(0)z = z$ for all $z \in Z = T_0O_G(0)$. Moreover the linear operators $L_s = D_x\Phi_s(0) : X \to X$ are isomorphisms since $L_sL_{s^{-1}} = \mathrm{id}_X$.

First we want to find a splitting $X = Z \oplus Y$ such that Y is invariant under all L_s. For this purpose take any projection $P : X \to Z$ with $Pz = z$ for all $z \in Z$. Using the *Haar measure* μ of the compact Lie group S [GSS88, Ch.XII.1c] we let

$$Qx = \int_{s \in S} PL_sx \, d\mu(s).$$

Then $Qz = z$ and $L_sQx = Qx$. Furthermore $Q^2x = \int_s \int_{\bar{s}} PL_sPL_{\bar{s}}x \, d\mu(\bar{s})d\mu(s)$, but since $PL_{\bar{s}}x \in Z$ and $PL_sz = z$ the integrand reduces to $PL_{\bar{s}}x$. Using the property $\int_S d\mu(s) = 1$ gives $Q^2x = Qx$ implying that Q is again a projection. Moreover $QL_{\bar{s}}x = \int_{s \in S} PL_{(s\bar{s})}x \, d\mu(s) = Qx = L_{\bar{s}}Qx$ by the invariance property of the Haar measure under translations. Hence, $X_1 = \mathrm{kernel}(Q)$ and $X_2 = \mathrm{kernel}(I - Q)$ are invariant under all L_s $(L_sX_j = X_j)$ and X_2 is even kept pointwise fixed, viz. $L_sx = L_sQx = Qx = x$ for all $x \in X_2$ and all $s \in S$. Obviously $Z \subset X_2$ and there is a closed subspace $\tilde{Y}$ such that $X_2 = Z \oplus \tilde{Y}$. Finally, let $Y = \tilde{Y} \oplus X_1$, then $X = Z \oplus Y$ and $L_sY = Y$ for all $s \in S$. Let $\tau_s = L_s|_Y$ be the restricted representation of S on Y.

Second we want to find a *slice* $\mathcal{Y}$ being invariant under Φ_s. A slice is a smooth submanifold of X with tangent space Y at $x_0 = 0$, such that, by transporting $\mathcal{Y}$ under the action of G, the whole neighborhood of $x_0 = 0$ can be foliated. We now define a diffeomorphism $\tilde{x} = M(x) = x + \mathcal{O}(\|x\|^2)$ such that the action induced by $\Phi_s(x)$ is given by the linear representation $L_s\tilde{x}$:

$$M(x) = \int_{s \in S} D_x\Phi_{s^{-1}}(0)\Phi_s(x) \, d\mu(s).$$

From $L_{s^{-1}}\Phi_s(\Phi_{\bar{s}}(x)) = L_{\bar{s}}L_{\bar{s}^{-1}}\Phi_{\bar{s}}(x)$, with $\tilde{s} = s\bar{s}$, the relation $M(\Phi_{\bar{s}}(x)) = L_{\bar{s}}M(x)$ follows. The inverse mapping N satisfies $N(L_s\tilde{x}) = \Phi_s(N(\tilde{x}))$. The slice is now the image of Y under N:

$$\mathcal{Y} = \{ N(y) \in X : y \in Y; \|y\| < \varepsilon \}.$$

Obviously $\mathcal{Y}$ is invariant under Φ_s as Y is so under L_s. We use $y \in Y$ as coordinates in $\mathcal{Y}$.

The transformation $(g, y) = \theta(x)$ is now obtained by solving the relation $\Phi_g(N(y)) = x$ by use of the implicit funtion theorem. Here the assumption (5.3c) is used. Denote the unique solution by $(g, y) = (\gamma(x), \eta(x))$; then the relation $\Phi_{sg_1}(x) = \Phi_{g_1g}(\Phi_s(N(y)) = \Phi_{g_1g}(N(\tau_sy))$ implies $\gamma(\Phi_{sg_1}(x)) = g_1\gamma(x)$ and $\eta(\Phi_{sg_1}(x)) = \tau_s\eta(x)$. This describes exactly the induced action given in (5.4). $\qquad\square$

Remark: We may assume that the linear operators τ_s, $s \in S$, are isometries on Y. Therefore we have to average the norm $\|\tau_s y\|$ over S with respect to the Haar measure. Then the new norm is equivalent to the old one, see [GSS88].

Henceforth we may assume that $\mathcal{X}$ is actually given by $G \times Y$ with the action $\Phi_{sh}(g, y) = (hg, \tau_s y)$. In particular, the invariance condition (5.2) is now equivalent to

$$H(g, \tau_s y) = H(y), \qquad J(hg, \tau_s y) = \begin{pmatrix} DL_h & 0 \\ 0 & \tau_s \end{pmatrix} J(g, y) \begin{pmatrix} DL_h^* & 0 \\ 0 & \tau_s^* \end{pmatrix}. \tag{5.5}$$

Thus, $J(g, y)$ is uniquely determined by $\overline{J}(y) = J(e, y)$. The vector field $X_H = JdH$ of a $\widetilde{G}$–invariant system now has the form

$$
\begin{aligned}
X_H(g, y) &= \begin{pmatrix} DL_h & 0 \\ 0 & I \end{pmatrix} \begin{pmatrix} \overline{J}_g(y) & \overline{J}_{gy}(y) \\ -\overline{J}_{gy}^*(y) & \overline{J}_y(y) \end{pmatrix} \begin{pmatrix} DL_h^* & 0 \\ 0 & I^* \end{pmatrix} \begin{pmatrix} 0 \\ \mathbf{d}_y H(y) \end{pmatrix} \\
&= \begin{pmatrix} DL_g \overline{J}_{gy}(y) \mathbf{d}_y H(y) \\ \overline{J}_y(y) \mathbf{d}_y H(y) \end{pmatrix},
\end{aligned}
$$

where $\overline{J}_g : \mathbf{g}^* \to \mathbf{g}$, $\overline{J}_{gy} : Y^* \to \mathbf{g}$, and $\overline{J}_y : Y^* \to Y$. In these coordinates (5.1) takes the form

$$\dot{g} = DL_g \overline{J}_{gy}(y) \mathbf{d}_y H(y). \tag{5.6}$$
$$\dot{y} = \overline{J}_y(y) \mathbf{d}_y H(y). \tag{5.7}$$

Note that the equation for g decouples in such a way that the y–equation can be solved first. Moreover, having solved (5.6) with some initial condition $g(t_0) = g_0$, the general solution is obtained by $hg(\cdot)$ with $h \in G$. It is an important point that (5.7) by itself is a generalized Hamiltonian system.

Proposition 5.2
The operator $\overline{J}_y$ defines a Poissson bracket $\{\cdot, \cdot\}_Y$ on Y.

Proof: Obviously $\overline{J}_y^* = -\overline{J}_y$. Moreover, Jacobi's identity for $\{\cdot, \cdot\}_Y$ follows from that on all of $\mathcal{X} = G \times Y$ by taking functions F, G, and H being independent of g: then $\{F, H\}_{\mathcal{X}} = \langle \begin{pmatrix} 0 \\ \mathbf{d}_y F \end{pmatrix}, J \begin{pmatrix} 0 \\ \mathbf{d}_y H \end{pmatrix} \rangle_{T\mathcal{X}} = \langle \mathbf{d}_y F, \overline{J}_y \mathbf{d}_y H \rangle_Y = \{F, H\}_Y$, and hence the assertion follows.   □

It will be seen later that even in the case of starting in the symplectic setting the (symmetry–) reduced system is no longer symplectic in general.

Now assume that (5.1) (resp. the system (5.6), (5.7)) has a finite–dimensional center manifold M_C and that Theorem 4.6 is applicable.

Theorem 5.3

*There is a S–invariant splitting $Y = Y_1 \times Y_2$, i.e. $\tau_s Y_i = Y_i$ for all $s \in S$, with $\dim Y_1 < \infty$
and a smooth function $h : U_1 \subset Y_1 \to Y_2$ with $h(0) = Dh(0) = 0$ and $h(\tau_s y_1) = \tau_s h(y_1)$,
$s \in S$, such that the set*

$$M_C = \{\, (g, y_1 + h(y_1)) \in G \times Y \ : \ g \in G, \ y_1 \in U_1 \subset Y_1 \,\}$$

*is a center manifold of system (5.6), (5.7). Moreover, the reduced Hamiltonian $\widetilde{H}(g, y_1)$
and the reduced Poisson structure $\widetilde{J}(g, y_1)$ are invariant under the reduced action action
$\widetilde{\Phi}$ of $\widetilde{G}$ given by $\widetilde{\Phi}_{sh}(g, y_1) = (hg, \tau_s y_1)$.*

Proof: Since $H(g, y) = H(y) = \frac{1}{2}\langle A_y y, y\rangle + \mathcal{O}(\|y\|^3)$, the linearization of the system
reads

$$\begin{pmatrix} \dot{\xi} \\ \dot{y} \end{pmatrix} = \begin{pmatrix} \overline{J}_g(0) & \overline{J}_{gy}(0) \\ -\overline{J}_{gy}^*(0) & \overline{J}_y(0) \end{pmatrix} \begin{pmatrix} 0 & 0 \\ 0 & A_y \end{pmatrix} \begin{pmatrix} \xi \\ y \end{pmatrix}$$

where $\xi \in \mathbf{g}$. Thus, $\mathbf{g}$ is contained in the kernel of the linear operator $K = JA$, and the
linear subspace X_1, being the tangent space at the center manifold, can be given as $\mathbf{g} \times Y_1$
with a finite–dimensional $Y_1 \subset Y$. Thus we have a center manifold

$$M_C = \{\, (g(\xi, y_1), y_1 + h(\xi, y_1)) \ : \ (\xi, y_1) \in V_1 \subset \mathbf{g} \times Y_1 \,\}$$

with a Poisson structure $\overline{J}(\xi, y_1)$.

Considering the equation on Y only, the linearization is just $\dot{y} = J_y(0) A_y y$, and thus
(5.7) has a center manifold spanned over the space Y_1, given by

$$\widehat{M_C} = \{\, y = y_1 + h(0, y_1) \ : \ y_1 \in U_1 \,\}.$$

The flow on the full center manifold is now given by the reduced system on Y_1 and the
g–equation (5.6), where $y = y_1 + h(0, y_1)$. However, as any initial data $g = g_0$ can be
chosen we have $M_C = G \times \widehat{M_C}$. The Poisson structure $\overline{J}(\xi, y_1)$ can be transformed into
the (g, y_1)–coordinates to obtain $\widetilde{J}(g, y_1)$.

It remains to be shown, that $\widetilde{J}$ is invariant under the action of G. Therefore recall
the special structure $M_C = G \times \widehat{M_C}$ and that the action $\Phi_h : \mathcal{X} \to \mathcal{X}$ of G acts only on
the g–component. Considering Φ_h as a diffeomorphism on $\mathcal{X}$ and using the fact that the
reduction process commutes with the application of diffeomorphisms we find the relation

$$\widetilde{J}(hg, y_1) = \begin{pmatrix} DL_h & 0 \\ 0 & I \end{pmatrix} \widetilde{J}(g, y_1) \begin{pmatrix} DL_h^* & 0 \\ 0 & I^* \end{pmatrix}.$$

Moreover, the invariance of the y_1–system under the S–action is a consequence of part
d) of the center manifold theorem 2.1 and the remark following the slice theorem 5.1.
Hence the result is proved. $\square$

Specializing this result to symplectic manifolds and using Theorem 4.1 gives

Corollary 5.4

Let the action Φ of $\widetilde{G} = G \odot S$ satisfy (5.3); let J in (5.1) be generated by a $\widetilde{G}$–invariant symplectic structure ω, and assume H is also $\widetilde{G}$–invariant. Then the reduced symplectic structure and the flow on the center manifold are again invariant under the (reduced) action.

5.3 Canonical Hamiltonian systems

A *canonical Hamiltonian system* is a system where the manifold $\mathcal{X}$ can be understood as a cotangent bundle T^*Q of a manifold Q modelled over some reflexive Banach space Q. Q is called the configuration space and T^*Q the phase space. Moreover, the symplectic structure has to be the canonical one. If convenient we denote the points in T^*Q by (q,p) where $p \in T_q^*Q$.

The importance of canonical Hamiltonian systems arises from the fact that they, and only they, may be transformed into a Lagrangian system. Lagrangian systems are always defined as differential equation on the tangent bundle TQ of a manifold. Via the fiber derivative (or the Legendre transform) we obtain an associated Hamiltonian system on T^*Q equipped with its canonical symplectic structure (see the next chapter).

Definition 5.5

The canonical one–form θ_{can} on T^*Q is uniquely defined, in local coordinates, by the relation

$$\theta_{(q,p)}(v,w) = \langle p,v \rangle_{T_qQ}$$

where $(v,w) \in Q \times Q^* = T_{(q,p)}T^*Q$.

The canonical two–form ω_{can} is given by $\omega_{can} = -\mathbf{d}\theta_{can}$.

As in the previous section we want to treat the case that the system is invariant under the action of a Lie group $\widetilde{G} = G \odot S$. In particular, we now assume that $\widetilde{G}$ acts via $\widehat{\Phi}$ on the configuration space Q directly. Hence, there is a *lifted action* Φ on the phase space T^*Q given by

$$\Phi_{\widetilde{g}}(q,p) = (\widehat{\Phi}_{\widetilde{g}}(q), (D_q\widehat{\Phi}_{\widetilde{g}}(q)^*)^{-1}p), \ \widetilde{g} \in \widetilde{G}.$$

This again is a severe restriction generating a lot of simplifications, but still general enough to treat many applications as those given below. Actions of this type are called point transformations; they preserve the canonical one– and two–form, since, letting $\overline{q} = \Phi_g(q)$, we have $\overline{v} = D_x\Phi_g(q)v$ for $v \in T_qQ$ and hence $\overline{\theta}_{(\overline{q},\overline{p})}(\overline{v},\overline{w}) = \langle \overline{p},\overline{v} \rangle = \langle (D\Phi_g(q)^*)^{-1}p, D\Phi_g(q)v \rangle = \langle p,v \rangle = \theta_{can(q,p)}(v,w)$.

Guided by Lemma 5.1 we can restrict ourself to the case that Q has the form $G \times \widetilde{Q}$ for some reflexive Banach space $\widetilde{Q}$ and that the action $\widehat{\Phi}$ is given by $\widehat{\Phi}_{sg_1}(g,\widetilde{q}) = (g_1g, \tau_s\widetilde{q})$ as in (5.4). We now introduce the new coordinates $\eta = DL_g^* w \in \mathbf{g}^*$, where $DL_g^* : T_g^* G \to \mathbf{g}^*$ is the adjoint of $DL_g : \mathbf{g} \to T_g G$. This defines an diffeomorphism between T^*G and $T°G = G \times \mathbf{g}^*$. Hence, $T^*(G \times \widetilde{Q})$ is diffeomorphic to $T°(G \times \widetilde{Q}) = G \times \mathbf{g}^* \times \widetilde{Q} \times \widetilde{Q}^*$. Letting $Y = \mathbf{g}^* \times \widetilde{Q} \times \widetilde{Q}^*$ leads us back to the situation considered above.

The diffeomorphism from $T^*(G \times \widetilde{Q})$ to $T°(G \times \widetilde{Q})$ induces the unique one–form $\theta°$ and the unique symplectic form $\omega°$ on $T°(G \times \widetilde{Q})$. They are given by

$$\theta°_{(g,\eta,q,p)}(v_1, w_1, x_1, y_1) = \langle \eta, DL_{g^{-1}} v_1 \rangle_\mathbf{g} + \langle p, x_1 \rangle_{Q_1},$$

$$\omega°_{(g,\eta,q,p)}((v_1, w_1, x_1, y_1), (v_2, w_2, x_2, y_2)) = \langle y_2, x_1 \rangle_{Q_1} - \langle y_1, x_2 \rangle_{Q_1} \tag{5.8}$$
$$+ \langle \eta, [DL_{g^{-1}} v_1, DL_{g^{-1}} v_2] \rangle_\mathbf{g} + \langle w_2, DL_{g^{-1}} v_1 \rangle_\mathbf{g} - \langle w_1, DL_{g^{-1}} v_2 \rangle_\mathbf{g},$$

where $(v_i, w_i, x_i, y_i) \in T_g G \times \mathbf{g}^* \times Q_1 \times Q_1^*$ and where $[\cdot, \cdot]$ is the Lie bracket on $\mathbf{g}$. For a derivation see [AM78, Thm.4.4.1] where $G \times Q_1$ has to be considered as a Lie group. The induced action Φ of $\widetilde{G}$ on $T°(G \times \widetilde{Q})$ becomes

$$\Phi_{sg_1}(g, \eta, \widetilde{q}, \widetilde{p}) = (g_1 g, \eta, \tau_s \widetilde{q}, \tau^*_{s^{-1}} \widetilde{p}). \tag{5.9}$$

Our aim is to show that center manifold systems of a canonical problem can always be put into canonical form. This is of course well known if no symmetries are present, since every finite–dimensional Hamiltonian system can be transformed into a canonical one by means of the classical Darboux theorem (see e.g. [AM78] or below). We generalize this theory by taking account of $\widetilde{G}$–invariance. Results for compact Lie groups are given in [GS84, Ch.II.27], but we need also the case of non–compact G.

Theorem 5.6

Let $\mathcal{X} = T°(G \times \widetilde{Q})$ *be equipped with* $\omega°$ *and let* $\widetilde{G} = G \odot S$ *act on* $\mathcal{X}$ *via* Φ *given in (5.9), where* S *is compact. Assume that the Hamiltonian* H *is* $\widetilde{G}$–invariant *and that the associated Hamiltonian system (5.1) has an equilibrium point* $x_0 = (e, \eta_0, 0, 0)$ *with a finite–dimensional center manifold* M_C. *Then in* M_C *there exist local coordinates* $(g, \eta, v, w) \in G \times \mathbf{g}^* \times Q_1 \times Q_1^*$, *with* Q_1 *being a finite–dimensional space, such that, in these coordinates, the reduced Hamiltonian system is* $\widetilde{G}$–invariant.

Proof: Employing the Theorems 4.1 and 5.3 we find that the reduced structure is non–degenerate and $\widetilde{G}$–invariant on a coordinate space $G \times Y_1$. It only remains to be shown that Y_1 has the form $\mathbf{g}^* \times T^* Q_1$.

Therefore consider the tangent space $T_{(e,0)} M_C = \mathbf{g} \times Y_1 \subset \mathbf{g} \times Y$ in the (g, y)–coordinates. However, we have

$$\omega°_{(e,0)}((\xi_1, 0), (\xi_2, 0)) = \omega_{can\, x_0}((D_g \Phi(\cdot, 0)[\xi_1], 0), (D_g \Phi(\cdot, 0)[\xi_2], 0)) = 0$$

for all $\xi_1, \xi_2 \in \mathbf{g}$, since $T_{q_0} O_G(q_0) \subset T_{(e,0)} Q$ is an isotropic subspace of $T_{x_0} T^* Q$. Hence, $\mathbf{g}$ is an isotropic subspace of $\mathbf{g} \times Y_1$.

This implies that $\dim(\mathbf{g}) \leq \dim(Y_1)$, and thus taking any basis $\{e_1, \ldots, e_n\}$ in $\mathbf{g}$ we may complete it to a *symplectic basis* $\{e_1, \ldots, e_{n+m}, \widetilde{e}_1, \ldots, \widetilde{e}_{n+m}\}$ of $\mathbf{g} \times Y_1$ according to the symplectic form $\omega^\circ_{(e,0)}$, i.e. we have the relations $\omega^\circ_{(e,0)}(e_i, \widetilde{e}_j) = \delta_{i-j}$ (Kronecker symbol), $\omega^\circ_{(e,0)}(e_i, e_j) = \omega^\circ_{(e,0)}(\widetilde{e}_i, \widetilde{e}_j) = 0$ for all i and j (see [AM78]). Hence, redefining the scalar product such that this basis is orthonormal, we have $Q_1 = \mathrm{span}\{e_{n+1}, \ldots, e_{n+m}\}$, $\mathbf{g}^* = \mathrm{span}\{\widetilde{e}_1, \ldots, \widetilde{e}_n\}$, and $Q_1^* = \mathrm{span}\{\widetilde{e}_{n+1}, \ldots, \widetilde{e}_{n+m}\}$. $\qquad\square$

Darboux's theorem is the classical tool to find canonical coordinates. We will use the *Lie transform method* of Moser and Weinstein [We71] which can easily be modified to fit our purposes. The essential new feature in our analysis is the invariance under the action of a Lie group $\widetilde{G}$. To retain this invariance we have to restrict the set of possible coordinates, namely to $\widetilde{G}$–invariant changes of the coodinates. A local diffeomorphism Θ is called $\widetilde{G}$–invariant if $\Theta(\Phi_{\widetilde{g}}(x)) = \Phi_{\widetilde{g}}(\Theta(x))$. For the special action Φ in (5.9) this implies

$$\Theta(g, \eta, \widetilde{q}, \widetilde{p}) = (g\Theta_1(\eta, \widetilde{q}, \widetilde{p}), \Theta_2(\eta, \widetilde{q}, \widetilde{p}), \Theta_3(\eta, \widetilde{q}, \widetilde{p}), \Theta_4(\eta, \widetilde{q}, \widetilde{p}))$$

where $\Theta_j(\eta, \tau_s \widetilde{q}, \tau^*_{s^{-1}} \widetilde{p}) = \Theta_j(\eta, \widetilde{q}, \widetilde{p})$ for $j = 1, 2$, $\Theta_3(\eta, \tau_s \widetilde{q}, \tau^*_{s^{-1}} \widetilde{p}) = \tau_s \Theta_3(\eta, \widetilde{q}, \widetilde{p})$, and $\Theta_4(\eta, \tau_s \widetilde{q}, \tau^*_{s^{-1}} \widetilde{p}) = \tau^*_{s^{-1}} \Theta_4(\eta, \widetilde{q}, \widetilde{p})$; for all $s \in S$.

In general, the $\widetilde{G}$–invariance of the original two–form will not be enough to guarantee the existence of $\widetilde{G}$–invariant coordinates. This is due to the non–compactness of G. For compact $\widetilde{G}$ the $\widetilde{G}$–invariant coordinates can easily obtained by averaging over $\widetilde{G}$ with respect to the Haar measure [GS84]. In the non–compact case we have to impose an additional condition on the two–form; however, for the center manifold applications we are dealing with, this condition always holds.

Theorem 5.7
The coordinates $(g, \eta, v, w) \in G \times \mathbf{g}^* \times Q_1 \times Q_1^*$ in Theorem 5.6 can be chosen such that the reduced Hamiltonian is $\widetilde{G}$–invariant and that the reduced symplectic structure is canonical (and hence G–invariant).

The proof of this theorem will be the content of the next section and is completed on page 55.

5.4 Darboux's theorem in the presence of symmetry

We would like to show that every $\widetilde{G}$–invariant symplectic form ω on $T^\circ(G \times Q_1) = G \times \mathbf{g}^* \times Q_1 \times Q_1^*$ can be transformed into ω° given in (5.8) by a $\widetilde{G}$–invariant change of coordinates. We remain in the case where Φ is given in (5.9).

A general one–form α and a general two–form β on $T^\circ(G \times Q_1)$ is $\widetilde{G}$–invariant if

$$\alpha_{(g,\eta,q,p)}(v_1, w_1, x_1, y_1) = \alpha_{(e,\eta,\tau_s q, \tau_{s-1}^* p)}(DL_{g^{-1}}v_1, w_1, \tau_s x_1, \tau_{s-1}^* y_1),$$
$$\omega_{(g,\eta,q,p)}((v_1, w_1, x_1, y_1), (v_2, w_2, x_2, y_2)) =$$
$$= \omega_{(e,\eta,\tau_s q, \tau_{s-1}^* p)}((DL_{g^{-1}}v_1, w_1, \tau_s x_1, \tau_{s-1}^* y_1), (DL_{g^{-1}}v_2, w_2, \tau_s x_2, \tau_{s-1}^* y_2))$$

for all $gs \in \widetilde{G} = G \odot S$. Obviously the canonical one– and two–forms are G–invariant.

To stay self–contained we now proof a helpful standard result involving Lie derivatives. We define the *pull back* of a two–form ω by a diffeomorphism Ψ to be the two–form $\Psi_* \omega$ given by

$$\Psi_* \omega_x(w_1, w_2) = \omega_{\Psi(x)}(D\Psi(x)[w_1], D\Psi(x)[w_2]).$$

Hence, $\Psi_* \omega$ is just the two–form induced from ω by the inverse mapping Ψ^{-1}. Using this definition the $\widetilde{G}$–invariance of a two–form β is simply given by $\Phi_{sg\,*}\beta = \beta$ for all $sg \in \widetilde{G}$. Now let v_t be a vector field and Ψ_t its flow, i.e. $\frac{d}{dt}\Psi_t(x) = v_t(\Psi_t(x))$; then the following result holds:

Lemma 5.8
Let ω_t be a smooth family of closed two–forms, then the formula

$$\frac{d}{dt}\Psi_{t\,*}\omega_t = \Psi_{t\,*}\left(\frac{d}{dt}\omega_t + \mathbf{d}\alpha_t\right) \tag{5.10}$$

holds, where α_t is the one–form defined by $\alpha_t(w) = \omega_t(v_t, w)$.

Remark: In the notations of the calculus on manifolds this is the consequence of the well–known relations $\frac{d}{dt}\Psi_{t\,*}\omega_t = \Psi_{t\,*}\mathcal{L}_{v_t}\omega_t$ and $\mathcal{L}_{v_t}\omega_t = \mathbf{i}_{v_t}\mathbf{d}\omega_t + \mathbf{d}\mathbf{i}_{v_t}\omega_t + \frac{d}{dt}\omega_t$, where $\mathcal{L}$ is the Lie derivative and $\mathbf{i}_v\beta(\cdot) = \beta(v, \cdot)$.

Proof: It is sufficient to prove the result in local coordinates. Then ω_t is generated by $\Omega_t(x)$ and $\mathbf{d}\omega = 0$ is equivalent to

$$\langle D\Omega_t[w_1]w_2, w_3 \rangle + \langle D\Omega_t[w_2]w_3, w_1 \rangle + \langle D\Omega_t[w_3]w_1, w_2 \rangle = 0$$

for all vector fields (compare with Lemma 4.5).

The left–hand side in (5.10) reads

$$\frac{d}{dt}\Psi_{t\,*}\omega_t(w_1, w_2) \quad = \quad \frac{d}{dt}\langle \Omega_t(\Psi_t(x))D\Psi_t(x)w_1(x), D\Psi_t(x)w_2(x)\rangle =$$
$$= \langle \frac{d}{dt}\Omega_t(\Psi_t(x))D\Psi_t(x)w_1(x), D\Psi_t(x)w_2(x)\rangle + \langle \Omega_t(\Psi_t)Dv_t[D\Psi_t w_1], D\Psi_t w_2\rangle$$
$$+ \langle D\Omega_t[v_t(\Psi_t)]D\Psi_t(x)w_1(x), D\Psi_t(x)w_2(x)\rangle + \langle \Omega_t(\Psi_t)D\Psi_t w_1, Dv_t[D\Psi_t w_2]\rangle.$$

Here we have used $\frac{d}{dt}D\Psi_t(x)w_i(x) = Dv_t(\Psi_t(x))[D\Psi_t(x)w_i(x)]$ which follows from the flow property of Ψ_t.

For the right–hand side we use $\alpha_t(\widetilde{w}) = \langle \Omega_t v_t, \widetilde{w} \rangle$ and $\mathbf{d}\omega_t = 0$ which results in

$$\mathbf{d}\alpha_t(\widetilde{w}_1, \widetilde{w}_2) = \langle D\Omega_t[\widetilde{w}_1]v_t + \Omega_t Dv_t[\widetilde{w}_1], \widetilde{w}_2 \rangle - \langle D\Omega_t[\widetilde{w}_2]v_t + \Omega_t Dv_t[\widetilde{w}_2], \widetilde{w}_1 \rangle$$
$$= \langle D\Omega_t[v_t]\widetilde{w}_1, \widetilde{w}_2 \rangle + \langle \Omega_t Dv_t[\widetilde{w}_1], \widetilde{w}_2 \rangle + \langle \Omega_t \widetilde{w}_1, Dv_t[\widetilde{w}_2] \rangle.$$

Letting $\widetilde{w}_i = D\Psi_t w_i$ and evaluating Ω_t and v_t at $\widetilde{x} = \Psi_t(x)$ exactly gives $\Psi_{t*}\mathbf{d}\alpha_t$; and the result follows. $\qquad\square$

A well–known consequence of this lemma is

Corollary 5.9

The flow mapping ϕ_t corresponding to the Hamiltonian vector field $J\mathbf{d}H$ preserves the symplectic form (i.e. $\phi_{t*}\omega = \omega$ for all t).

Proof: Applying Lemma 5.8 gives $\frac{d}{dt}\phi_{t*}\omega = \phi_{t*}(\frac{d}{dt}\omega + \mathbf{d}\alpha_t)$ where $\alpha_t(z) = \omega(J\mathbf{d}H, z) = \langle \mathbf{d}H, z \rangle$. Now $\frac{d}{dt}\omega = 0$ and $\mathbf{dd}H = 0$ completes the proof. $\qquad\square$

First we give a version of Poincaré's lemma which shows that every closed two–form ω ($\mathbf{d}\omega = 0$) is locally exact, i.e. it is the exterior derivative of a one–form α ($\mathbf{d}\alpha = \omega$). Again, we generalize the classical result to the $\widetilde{G}$–invariant case.

Theorem 5.10
(Poincaré's Lemma)

Let ω be a closed $\widetilde{G}$–invariant two–form on $T^\circ(G \times Q_1)$. Assume that $\omega_{(e,0,0,0)}$ restricted to $T_e G \times \{0\}$ vanishes completely, then there is a $\widetilde{G}$–invariant one–form α such that $\mathbf{d}\alpha = \omega$.

Remark: In the case of $\widetilde{G}$ being non–compact we cannot expect to find a $\widetilde{G}$–invariant α for every $\widetilde{G}$–invariant ω. This is easily seen by taking $\widetilde{G} = G = \mathbb{R}$, $Q_1 = \{0\}$, and $\omega_{(x,y)}((v_1, w_1), (v_2, w_2)) = v_1 w_2 - v_2 w_1$. Since ω is independent of x it is $\widetilde{G}$–invariant. But α has to have the form $\alpha_{(x,y)}(v, w) = xw + \mathbf{d}f(v, w)$ for some function f. Hence α cannot be $\widetilde{G}$–invariant.

However, it will be seen in the proof that the condition given in the theorem is not necessary. It can be weakened as follows: There is a $\widetilde{\eta} \in \mathbf{g}^*$ such that

$$\omega_{(e,0,0,0)}((\xi_1, 0, 0, 0), (\xi_2, 0, 0, 0)) = \langle \widetilde{\eta}, [\xi_1, \xi_2] \rangle_{\mathbf{g}}.$$

Since the two–form defined by the right–hand side is the derivative of the G–invariant one–form $\alpha^{\widetilde{\eta}}$ given by $\alpha^{\widetilde{\eta}}_{(g,\eta)}(v, w) = \langle \widetilde{\eta}, DL_{g^{-1}}v \rangle_{\mathbf{g}}$ the generalization is immediate.

Proof: Analogous to the classical proof we define a family of diffeomorphisms Ψ_ρ, $\rho > 0$, on $T^\circ(G \times Q_1)$ by $\Psi_\rho(g, \eta, q, p) = (g, \rho\eta, \rho q, \rho p)$. Note that the g–variable cannot be contracted, since generally G is not a linear space and since we have to retain the symmetry. Obviously, Ψ_ρ is the flow generated by the vector field $v_\rho(g, \eta, q, p) = \frac{1}{\rho}(0, \eta, q, p)$.

Using Lemma 5.8 yields $\frac{d}{d\rho}\Psi_{\rho\ast}\omega = \Psi_{\rho\ast}\mathbf{d}\widetilde{\alpha}_\rho$ with $\widetilde{\alpha}_\rho(w) = \omega(v_\rho, w)$. Set $\alpha_\rho(w) = \Psi_{\rho\ast}\widetilde{\alpha}(w) = \widetilde{\alpha}_{\Psi_\rho(x)}(D\Psi_\rho w)$ then $\frac{d}{d\rho}\Psi_{\rho\ast}\omega = \mathbf{d}\alpha_\rho$, since pull back and exterior derivative commute. Hence, letting $\alpha = \int_0^1 \alpha_\rho\, d\rho$ we obtain $\mathbf{d}\alpha = \int_0^1 \mathbf{d}\alpha_\rho\, d\rho = \int_0^1 \frac{d}{d\rho}\Psi_{\rho\ast}\omega\, d\rho = \Psi_{1\ast}\omega - \Psi_{0\ast}\omega$. However, $\Psi_{1\ast}\omega = \omega$ and $\Psi_{0\ast}\omega = 0$ by the assumption that ω restricted to g vanishes.

Moreover, we see that α is $\widetilde{G}$–invariant by construction: Notice that Ψ_ρ is $\widetilde{G}$–invariant since it is linear in the (η, q, p)–variable and the identity on G. The $\widetilde{G}$–invariance of ω now implies that $\widetilde{\alpha}_\rho$, and hence α_ρ and α, are $\widetilde{G}$–invariant also. $\qquad\square$

A closely related result, which is due to [We71], is concerned with a generalization of Poincaré's Lemma in the presence of a submanifold.

Theorem 5.11

Let ω be a symplectic form on $\mathcal{X}$, which vanishes on a submanifold $\mathcal{M}$, i.e. $\omega_m(v_1, v_2) = 0$ for all $m \in \mathcal{M}$ and $v_1, v_2 \in T_m\mathcal{X}$. Then there exists a one–form α with $\mathbf{d}\alpha = \omega$ vanishing on $\mathcal{M}$ ($\alpha_m(v) = 0$ for all $m \in \mathcal{M}$ and $v \in T_m\mathcal{X}$).

Proof: We choose local coordinates such that $\mathcal{M}$ is given as a linear subspace $M \subset X$. Now we proceed as in the above proof and define Ψ_ρ as a contraction in the complementary subspace X_2 only ($X = M \oplus X_2$). To see that the one–form α constructed as above vanishes on M, we note that $\Psi_\rho(m) = m$ for all ρ and $m \in M$. Hence, $\omega_m = 0$ implies $\widetilde{\alpha}_{\rho m} = 0$ and $\alpha_{\rho m} = 0$. Integration over ρ gives the result. $\qquad\square$

Using this Poincaré's Lemma we can now proceed to prove Darboux's theorem. To retain the $\widetilde{G}$–invariance we only consider $\widetilde{G}$–invariant changes of the coordinate system. Of course, the pull back $\Psi_\ast\omega$ of a $\widetilde{G}$–invariant two–form ω by a $\widetilde{G}$–invariant diffeomorphism Ψ is again $\widetilde{G}$–invariant.

Theorem 5.12

(Darboux's Theorem)

Let ω be a $\widetilde{G}$–invariant symplectic form on $T^\circ(G \times Q_1)$ such that g is an isotropic subspace of $T_{(e,0)}T^\circ(G \times Q_1)$. Then there is a $\widetilde{G}$–invariant change of coordinates Ψ such that the symplectic form $\Psi_\ast\omega$ is canonical.

Proof: We again stay close to the classical proof in [We71, AM78] using the Lie transform method. Let ω_1 be the unique $\widetilde{G}$–invariant symplectic form on $T^\circ(G \times Q_1)$ defined by

$$\omega^1_{(e,\eta,q,p)}((v_1, w_1, x_1, y_1), (v_2, w_2, x_2, y_2)) = \omega_{(e,0,0,0)}((v_1, w_1, x_1, y_1), (v_2, w_2, x_2, y_2))$$

and let $\omega^0 = \omega^1 - \omega$ and $\omega_\rho = \omega + \rho\omega^0$. Hence, ω^0, ω^1, and ω_ρ are closed and $\widetilde{G}$–invariant. From $\omega^0_{(g,0,0,0)} = 0$ we obtain the existence of a neighborhood $U_1 \subset Y_1$ such that ω_ρ is non–degenerate on $G \times U_1$ for all $\rho \in [0,1]$. Moreover, according to the previous theorem there is a $\widetilde{G}$–invariant one–form α^0 with $\mathbf{d}\alpha^0 = \omega^0$. Defining the vector field v_ρ on $T^\circ(G \times Q_1)$ by $\omega_\rho(v_\rho, w) = -\alpha^0(w)$ we obtain, maybe on a smaller neighborhood $\widetilde{U}_1$, the flow operator Ψ_ρ, $\rho \in [0,1]$, with $\Psi_0 =$ identity. Then by Lemma 5.8 we obtain

$$\tfrac{d}{d\rho}\Psi_{\rho*}\omega_\rho = \Psi_{\rho*}(\tfrac{d}{d\rho}\omega_\rho - \mathbf{d}\alpha^0) = \Psi_{\rho*}(\omega^0 - \mathbf{d}\alpha^0) = 0.$$

Hence, $\Psi_{1*}\omega_1 = \Psi_{0*}\omega_0 = \omega$ and thus Ψ_1^{-1} transforms $\omega_0 = \omega$ into $\omega_1 = \omega^1$. Moreover, the mapping Ψ and its inverse are $\widetilde{G}$–invariant due to the invariance of v_ρ.

It remains to be shown that the simplified two–form ω^1 can be put into canonical form. Therefore it is sufficient to consider the tangent space at $(e,0,0,0)$. However, as in the last part of the proof of Theorem 5.6, canonical coordinates (respectively a symplectic basis) can be found. Here we use the assumption that $\mathbf{g}$ is an isotropic subspace. □

We now give the

Proof of Theorem 5.7:
In order to apply Theorem 5.12 we only have to know that $\mathbf{g}$ is an isotropic subspace. However, this was already shown in the proof of Theorem 5.6. □

For the flattening of center manifolds (Section 4.3) and for the relaxed natural reduction of Lagrangian systems (Section 6.5) we will need the following generalization of Darboux's theorem in the presence of a submanifold. The result is due to Weinstein [We71]. The proof is completely analogous to that of Theorem 5.12. However, we employ Poincaré's lemma in the version of Theorem 5.11.

Lemma 5.13
Let M be a submanifold of $\mathcal{X}$. Let ω_0 and ω_1 be (strongly non–degenerate) symplectic forms on $\mathcal{X}$ such that $\omega_0 - \omega_1$ vanishes on M. Then, around each $m \in M$ there exists a local diffeomorphism Ψ which keeps M pointwise fixed and transforms ω_0 into ω_1, i.e. $\Psi_*\omega_1 = \omega_0$ in a whole neighborhood of $m \in \mathcal{X}$.

As a consequence we obtain the following general result which was already used in Section 4.3 in the case of center manifolds.

Theorem 5.14
Let $\mathcal{X}$ be a symplectic manifold modelled over a Hilbert space H and let M be a symplectic submanifold of $\mathcal{X}$. Then, around each $m \in M$ there are local canonical coordinates $(q_1, q_2, p_1, p_2) \in H$ such that M is given by $(q_2, p_2) \equiv 0$.

Proof: Without loss of generality we can work in a local coordinate system and assume $m = 0$. We let $X_1 = T_m\mathcal{M}$ and $X_2 = X_1^\perp$ is the orthogonal complement in H. Then X_1 and X_2 are symplectic Hilbert spaces with canonical coordinates $(\overline{x}_1, \overline{x}_2) = (\overline{q}_1, \overline{p}_1, \overline{q}_2, \overline{p}_2) \in X_1 \times X_2 = H$ ([We71]). We write $\overline{\Omega} = \begin{pmatrix} \overline{\Omega}_1 & 0 \\ 0 & \overline{\Omega}_2 \end{pmatrix}$ for the canonical structure. The submanifold is then given by $\overline{x}_2 = h(\overline{x}_1) = \mathcal{O}(\|\overline{x}_1\|^2)$.

To flatten the submanifold we let $(\widehat{x}_1, \widehat{x}_2) = (\overline{x}_1, \overline{x}_2 - h(\overline{x}_1))$ and obtain the induced symplectic form

$$\widehat{\Omega}(\widehat{x}) = \begin{pmatrix} \overline{\Omega}_1 + Dh(\widehat{x}_1)^T\overline{\Omega}_2 Dh(\widehat{x}_1) & Dh(\widehat{x}_1)^T\overline{\Omega}_2 \\ \overline{\Omega}_2 Dh(\widehat{x}_1) & \overline{\Omega} \end{pmatrix}.$$

In order to apply the previous lemma we make a further change of variables $(\widehat{x}_1, \widehat{x}_2) = (R(\widetilde{x}_1, \widetilde{x}_2), S(\widetilde{x}_1)\widetilde{x}_2)$, such that the induced symplectic structure $\widetilde{\Omega}$ is canonical on $\widetilde{x}_2 = \widehat{x}_2 = 0$. We find

$$
\begin{aligned}
\widetilde{\Omega}(\widetilde{x}_1, \widetilde{x}_2) &= \begin{pmatrix} \widetilde{\Omega}_1 & \widetilde{\Omega}_{12} \\ -\widetilde{\Omega}_{12}^* & \widetilde{\Omega}_2 \end{pmatrix} \quad \text{with} \\
\widetilde{\Omega}_1(\widetilde{x}_1, 0) &= D_1 R^T\left[\overline{\Omega}_1 + Dh(R)^T\overline{\Omega}_2 Dh(R)\right] D_1 R, \\
\widetilde{\Omega}_{12}(\widetilde{x}_1, 0) &= D_1 R^T\left[\overline{\Omega}_1 D_2 R - Dh(R)^T\overline{\Omega}_2 S\right], \\
\widetilde{\Omega}_2(\widetilde{x}_1, 0) &= S^T\overline{\Omega}_2 S - S^T\overline{\Omega}_2 Dh(R)D_2 R - D_2 R^T Dh(R)^T\overline{\Omega}_2 S + D_2 R^T\overline{\Omega}_1 D_2 R.
\end{aligned}
$$

Here R and $D_j R = D_{\widetilde{x}_j}R$ are evaluated at $(\widetilde{x}_1, 0)$ and S at $\widetilde{x}_1$.

Since $\widetilde{\Omega}_1$ defines a symplectic form on $\mathcal{M}$ ($\widetilde{x}_2 = 0$), the classical Darboux theorem provides a transformation $\widetilde{R} : \widetilde{x}_1 \to \widehat{x}_1 = R(\widetilde{x}_1, 0)$, such that $\widetilde{\Omega}_1(\widetilde{x}_1, 0) \equiv \Omega_{1\,can}$. Next we see that $D_2 R(\widetilde{x}_1, 0)$ can be chosen as $\overline{\Omega}_1^{-1} Dh(R)^T\overline{\Omega}_2 S$ implying $\widetilde{\Omega}_{12}(\widetilde{x}_1, 0) \equiv 0$. Inserting this into the expression for $\widetilde{\Omega}_2$ we find, after a cancellation,

$$\widetilde{\Omega}_2(\widetilde{x}_1, 0) = S^T(\overline{\Omega}_2 - \overline{\Omega}_2 Dh(R)\overline{\Omega}_1^{-1} Dh(R)^T\overline{\Omega}_2)S.$$

Obviously a suitable choice of S yields $\widetilde{\Omega}_2(\widetilde{x}_1, 0) \equiv \Omega_{2\,can}$.

Altogether, we now have $\widetilde{\Omega}(\widetilde{x}_1, 0) = \Omega_{can}$, and hence Lemma 5.13 provides us with a local diffeomorphism $\widetilde{\Psi}$ which keeps each point $(\widetilde{x}_1, 0)$ fixed and which transforms $\widetilde{\Omega}$ into Ω_{can}. The desired coordinate system is now given by $(q_1, p_1, q_2, p_2) = (x_1, x_2) = \widetilde{\Psi}(\widetilde{x}_1, \widetilde{x}_2)$.
$\square$

Remark: For manifolds modelled over a reflexive Banach space a similar result is possible. We find local coordinates $(x_1, x_2) \in X_1 \oplus X_2$ such that the submanifold is given by $x_2 \equiv 0$, that the symplectic form is constant, and that X_1 and X_2 are symplectically orthogonal.

Additionally we need a result concerning Darboux's theorem showing us how small the corrections on the coordinate system can be chosen if the original symplectic form is already close to the canonical one. This result complements Theorem 4.2 and Corollary 4.3, where already good approximations for a canonical form were found.

Theorem 5.15
Assume that $\Omega(q,p) = \begin{pmatrix} 0 & I \\ -I & 0 \end{pmatrix} + \mathcal{O}(\|(q,p)\|^m)$, with $m \geq 1$, generates a symplectic form. Then there are canonical coordinates $(\bar{q},\bar{p})$ such that $(\bar{q},\bar{p}) = (q,p) + \mathcal{O}(\|(q,p)\|^{m+1})$.

Proof: We just go through the proof of Theorem 5.12 once again with the trivial Lie group $G = \{e\}$. Note that ω^0 vanishes in the origin of order m. Hence, the one–form α^0, constructed as in the proof of Theorem 5.10, vanishes of order $m+1$, and the same is true for the vector field v_t. By classical results on ordinary differential equations we obtain $(\bar{q},\bar{p}) = \Psi_1(q,p) = (q,p) + \mathcal{O}(\|(q,p)\|^{m+1})$. $\square$

Specifying this result to the case in Theorem 4.2 and Corollary 4.3, where $m = 2n-4$, we obtain that the new Hamiltonian $\overline{H}(\bar{q},\bar{p})$ satisfies

$$\overline{H}(\bar{q},\bar{p}) = H(\bar{q},\bar{p}) + \mathcal{O}(\|(\bar{q},\bar{p})\|^{2n-2}), \tag{5.11}$$

since the corrections enter in the Hamiltonian at order $2(2n-3)$ which is larger than or equal to $2n-2$ as $n \geq 2$.

5.5 Reversible systems

A Hamiltonian system is called reversible, if there is a reflection in phase space such that reflecting a solution and reversing its time direction again gives a solution. For instance, in canonical systems with $H = H(q,p)$ this is always the case when H is an even function of p. Then having a solution $(q(t),p(t))$ we find another solution by reversing time and changing the sign of p, viz. $(q(-t),-p(-t))$ is again a solution. In the elliptic problems on cylindrical domains considered in Part II reversiblility is a consequence of a reflection symmetry with respect to the axial variable.

This concept can be made abstract by the following definition. Let $\mathcal{X}$ be a manifold with symplectic (or Poisson) structure $J : T^*\mathcal{X} \to T\mathcal{X}$ and a Hamiltonian H. Then the Hamiltonian system is called *reversible*, if there exists a smooth diffeomorphism $\mathcal{R} : \mathcal{X} \to \mathcal{X}$ such that

$$\mathcal{R} \circ \mathcal{R} = \text{Id}_\mathcal{X}, \quad H(\mathcal{R}(x)) = H(x), \quad D\mathcal{R}(x)J(x)D\mathcal{R}^*(x) = -J(\mathcal{R}(x)) \tag{5.12}$$

for all $x \in \mathcal{X}$. Note the minus sign in front of the J; it is responsible for the time reversal. In [MRS90] $\mathcal{R}$ is also called an antisymplectic diffeomorphism. Again we find that $\mathcal{R}(x(-t))$ is a solution whenever $x(t)$ is one.

Having a Hamiltonian system being invariant under the action Φ of a Lie group $\widetilde{G}$ which is also reversible, we realize that there is a second action $\Phi^{\mathcal{R}}$ on the system defined by

$$\Phi_g^{\mathcal{R}} = \mathcal{R} \circ \Phi_{\tilde{g}} \circ \mathcal{R}.$$

In most cases this action will be expressible through the old action but with shifted arguments. Hence, we assume that there is a smooth group automorphism $\mathcal{T} : \widetilde{G} \to \widetilde{G}$ (i.e. $\mathcal{T}(\tilde{g}_1\tilde{g}_2) = \mathcal{T}(\tilde{g}_1)\mathcal{T}(\tilde{g}_2)$) such that

$$\Phi_g^{\mathcal{R}} = \Phi_{\mathcal{T}(\tilde{g})}. \tag{5.13}$$

Example: Once again we study the beam problem of Chapter 11 (see also the example on page 44) where the Lie group $\widetilde{G} = G$ is the group of Euclidian transformations on $\mathbb{R}^3$ acting on the function space $Q = L_2(\Sigma)^3$ through

$$\widehat{\Phi}_{(R,r)} : \begin{pmatrix} u_1 \\ u_2 \\ u_3 \end{pmatrix} \to \begin{pmatrix} r_1 \\ r_2 \\ r_3 \end{pmatrix} + R \begin{pmatrix} u_1 \\ u_2 \\ u_3 \end{pmatrix},$$

where u_j are functions over Σ and in $(R,r) \in SO(3) \times \mathbb{R}^3 = G$ the matrix R represents the rotation and r the translation. On $\mathcal{X} = T^*Q = L_2(\Sigma)^6$ this defines the action $\Phi_{(R,r)}(u,v) = (r + Ru, Rv)$ since $R^{-1*} = R$.

The reversibility mapping $\mathcal{R}$ is given by $\mathcal{R}(u,v) = (\Gamma u, -\Gamma v)$ where $\Gamma u = (u_1, u_2, -u_3)$. The action $\Phi^{\mathcal{R}}$ becomes

$$\Phi_{(R,r)}^{\mathcal{R}}(u,v) = (\Gamma(r + R\Gamma u), -\Gamma R(-\Gamma v));$$

and hence the automorphism $\mathcal{T}$ takes the form $\mathcal{T}(R,r) = (\Gamma R \Gamma, \Gamma r)$.

The theory developed for $\widetilde{G}$–invariant systems can now be generalized to reversible systems. Therefore we assume that the automorphism $\mathcal{T}$ respects the direct splitting of $\widetilde{G}$ into $G \odot S$ as given in (5.3). This means

$$\mathcal{T}(gs) = \mathcal{T}_G(g)\mathcal{T}_S(s) \qquad \text{for all } gs \in \widetilde{G}. \tag{5.14}$$

Here $\mathcal{T}_G$ and $\mathcal{T}_S$ are the restrictions of $\mathcal{T}$.

Starting from an equilibrium $x_0 \in \mathcal{X}$ with $\mathcal{R}(x_0) = x_0$ we follow the constructions of the previous sections we see that at each step the reversibility can be retained. In particular in the slice theorem 5.1 we find a slice $\mathcal{Y}$ with $\mathcal{R}(\mathcal{Y}) = \mathcal{Y}$ with coordinates $y \in Y$ such that the action of $\mathcal{R}$ on Y can be assumed to be linear and isometric. Then the center manifold theory guarantees the existence of a center manifold M_C being again reversible. Moreover, both, Poincaré's Lemma and Darboux's theorem respects reversibility, since the natural constructions in the proofs do not destroy the reversibility.

Theorem 5.16

Let $\widetilde{G} = G \odot S$ satisfy (5.3) and let $\mathcal{X} = T^{\circ}(G \times \widetilde{Q})$ be equipped with the canonical form ω°. Assume that the Hamiltonian H is $\widetilde{G}$-invariant and has a relative equilibrium point $(\eta_0, 0, 0) \in \mathbf{g}^* \times T^*\widetilde{Q}$, with a finite-dimensional center manifold M_C. If, additionally, the system is reversible with respect to $\mathcal{R}$ such that (5.13), (5.14), and $\mathcal{R}(g, \eta_0, 0, 0) = (\mathcal{T}_G(g), \eta_0, 0, 0)$ hold; then the reduced Hamiltonian system can be written in canonical coordinates in $T^{\circ}(G \times Q_1)$ such that it is $\widetilde{G}$-invariant and reversible.

Chapter 6

Lagrangian systems

In classical mechanics most Hamiltonian systems are derived from a Lagrangian system via the so–called Legendre transformation. However it is the Hamiltonian theory which attracted more attention because of its greater flexibility. For instance, the class of Hamiltonian systems is closed under the action of diffeomorphisms. Moreover, there is a well–established theory of canonical transformations leaving the canonical symplectic structure invariant. For Lagrangian system no corresponding transformation rules are known. In fact, the best way to find transformations is to do canonical transformations in the associated Hamiltonian system and then transfer the result back into a Lagrangian one, being equivalent to the original one. While in Hamiltonian systems the new Hamiltonian is just the composition of the old one with the transformation, there is no simple relation between the new and the old Lagrange function.

Similar difficulties arise when reducing a Lagrangian system onto its center manifold. It turns out that there exist linear Lagrangian systems for which the flow on the center manifold cannot be generated by a Lagrangian system. But even if it is possible, only the detour over the associated Hamiltonian system and its center manifold reduction will enable us to construct this reduced Lagrangian on the center manifold. In contrast to the Hamiltonian case, this Lagrangian is not unique, since choosing different canonical coordinates on the center manifold will generate different Lagrange functions.

First we give a short outline of known reduction methods for variational problems, the Lyapunov–Schmidt reduction method and the projection method. The first provides an exact procedure to reduce 'steady' problems (i.e. $\dot{q} = 0$) to simpler ones. The projection method is appropriate for Lagrangian problems, but, in its simplest form, has no mathematical foundation and often gives wrong results.

Second we introduce the abstract notations of infinite–dimensional Lagrangian systems and their relation to canonical Hamiltonian systems. We follow the ideas in [CM74] and [HM83, Ch.5.3]. In Section 6.3 we then treat linear systems and establish a necessary and sufficient condition to decide whether a linear finite–dimensional Hamiltonian system can be transformed into a Lagrangian system by suitably chosen canonical coordinates. The result is that this is possible if and only if the dimension of the kernel of the associated linear operator K is not larger than half the dimension of the full space. This simple criterion involving only spectral properties is especially useful in cases where the calculation of the reduced Hamiltonian is difficult.

This linear theory carries over to nonlinear systems locally on the center manifold. We then try to compare our reduction method with the projection method and pose the *problem of natural reduction*. This means that the center manifold reduction and the projection method yield the same result. On the one hand this would provide a justification of the projection method and on the other hand it selects from all the possible reduced Lagrangian systems a physically reasonable one, see Section 11.4 for an example. Unfortunately, the mathematical tools are not yet developed to decide whether natural reduction is always possible. We are only able to solve the *relaxed natural reduction problem* by using the theory of flattening of center manifolds.

In Sections 6.6, 6.7 we consider systems being invariant under the action of a Lie group G. In particular, we study the structure of center manifolds near *relative equilibria*. These are curves in the Lie group which appear, after factoring out the symmetry, as an equilibrium in the reduced system. We again derive conditions under which the flow on a corresponding center manifold is generated by a reduced Lagrangian. Thus, we are able to study all solutions staying close to the relative equilibrium for all time.

6.1   Lyapunov–Schmidt and projection method

Reduction of variational problems to simpler variational problems is well–known in the context of the Lyapunov–Schmidt reduction procedure. This applies, compared to our theory, only to steady problems. Classical examples are elliptic systems on bounded domains [CH82] or periodic solutions in Hamiltonian systems [MRS90] which are treated as stationary points of the action integral in the space of all periodic functions.

The general setting is given in [Ki88]. For a potential function $V : Q \to \mathbb{R}$, where Q is a Hilbert space, we assume that $q = 0$ is a stationary point of V, i.e. $DV(0) = 0$. Moreover, we assume that Q splits into $Q_1 \oplus Q_2$ such that V has an expansion

$$V(q_1 + q_2) = \tfrac{1}{2}\langle A_1 q_1, q_1 \rangle + \tfrac{1}{2}\langle A_2 q_2, q_2 \rangle + R(q_1, q_2), \quad \text{with } R = \mathcal{O}(\|q\|^3).$$

The splitting is chosen such that $A_2 : D(A_2) \subset Q_2 \to Q_2$ has a bounded inverse A_2^{-1}. In most applications, Q_1 is finite–dimensional and $A_1 = 0$, but this is not necessary at all, see [Mi91b] for an example with $\dim Q_1 = \infty$.

To find all possible stationary points close to $q = 0$ we study the Euler–Lagrange equations

$$G_1(q_1, q_2) = A_1 q_1 + D_{q_1} R(q_1, q_2) = 0,$$
$$G_2(q_1, q_2) = A_2 q_2 + D_{q_2} R(q_1, q_2) = 0.$$

Because of the invertibility of A_2 the second equation can be solved locally for $q_2 = g(q_1)$, under suitable smoothness conditions on R. Thus, it remains to study the reduced problem

$$\widetilde{G}(q_1) = G(q_1, g(q_1)) = A_1 q_1 + D_{q_1} R(q_1, g(q_1)) = 0,$$

which is usually called the bifurcation equation.

The important feature is that this reduced problem corresponds to the reduced variational problem given by the reduced potential $\widetilde{V}(q_1) = V(q_1 + g(q_1))$. This means we have $D_{q_1} \widetilde{V}(q_1) = \widetilde{G}(q_1)$. This follows from $D_{q_1} \widetilde{V}[\rho] = D_{q_1} V[\rho] + D_{q_2} V[D_{q_1} g[\rho]]$ and the fact that g was determined such that $D_{q_2} V(q_1 + g(q_1))[\cdot] = 0$. This provides a powerful tool in local bifurcation theory for variational problems, since the reduced variational problem for $\widetilde{V}$ on Q_1 is much simpler than the original one, see e.g. [CH82, Ch.4.11] and [Ki88].

For variational problems of the above type it is often advantageous to consider the associated *gradient flow* $\dot{q} = -DV(q)$. Then, V is a Lyapunov function (i.e. $\dot{V} \leq 0$) which is only constant on equilibria. To this dynamical system the center manifold theory is again applicable giving a reduced differential equation on Q_1:

$$\dot{q}_1 = A_1 q_1 + D_{q_1} R(q_1, h(q_1)).$$

In general, h is different from the function g of the Lyapunov–Schmidt reduction, and the reduced problem is not a gradient system with respect to the reduced potentials $\widetilde{V}$ or $\widehat{V}(q_1) = V(q_1 + h(q_1))$. However, in view of Theorem 4.4 we have

$$\dot{q}_1 = -\widehat{J}(q_1) D\widehat{V}(q_1) \qquad \text{with } \widehat{J}(q_1) = I_{Q_1} + Dh(q_1)^T Dh(q_1).$$

Here $\widehat{J}$ defines the induced metric on the center manifold and the reduced equation is a gradient system with respect to this metric.

For the problem of periodic solutions in Hamiltonian systems the variational problem is defined on a loop space of continuous differentiable functions which is a non–reflexive Banach space. There, a more sophisticated reduction method, called the *splitting lemma of Magnus*, is needed, see [MRS90].

For the Lagrangian problems $L = L(q, \dot{q})$ the Lyapunov–Schmidt reduction can only deal with the steady part, viz. $V(q) = L(q, 0)$. However, we are interested in a family of solutions which does not have a prescribed time–dependence like (quasi–) periodic or homoclinic. Especially, we do not want to restrict the analysis to certain function classes in which the variational problem for $\int L(q(t), \dot{q}(t)) \, dt$ could be posed. In fact, it would be very helpful in this context if a precise variational characterization of all solutions on the center manifold could be found.

The applications of our theory are mainly in the field of elliptic partial differential equations in cylindrical domains where the axial variable plays the role of time. Typical examples appear in hydrodynamics when flows through channels are considered. In elasticity, the steady deformations of cylindrical bodies are relevant and lead to the rod theories. We will deal with problems of this type in Part II.

In engineering and applied sciences, reductions of this type are in use for many years. These methods are called *projection methods* or *Galerkin methods*. They are formally developed in the context of rod theory in [An72]. However, they are only approximative, sometimes even wrong in the leading terms (see the discussions at the end of Section 10.4 and Section 11.5). Nevertheless, they often give the right qualitative behavior. It is our goal to give a mathematically rigorous basis for these methods. For problems in rod theory this will be achieved in Chapters 10 and 11.

Projection method:

Given is a Lagrangian $L(q, \dot{q})$ on a space of high dimension. The solutions under consideration (e.g. those on the center manifold) are shown (or assumed) to be given by

$$q = r(\tilde{q}, \dot{\tilde{q}}), \quad \dot{q} = s(\tilde{q}, \dot{\tilde{q}}) \tag{6.1}$$

where the variable $\tilde{q}$ lies in a space of much smaller dimension. The projected Lagrangian L_P is then defined as

$$L_P(\tilde{q}, \dot{\tilde{q}}) = L(r(\tilde{q}, \dot{\tilde{q}}), s(\tilde{q}, \dot{\tilde{q}})). \tag{6.2}$$

Then the Lagrangian system corresponding to L_P is studied in order to analyse the solutions $\tilde{q}(t)$.

Yet, in general, inserting such a solution $\tilde{q}$ in (6.1) will not generate a solution $q(t)$ of the original Lagrangian system. However, in the case $q = r(\tilde{q})$ and $\dot{q} = Dr(\tilde{q})\dot{\tilde{q}}$ the ansatz (6.1) can be considered as a constraint manifold. Accepting this constraint the projection method yields a consistent reduced functional L_P. To be able to return to the full problem one has to find those constraints which are naturally satisfied by the solutions of the full (unconstrained) problem.

The center manifold reduction presented above picks out an exact solution manifold for the full problem and provides a reduced Lagrangian $\overline{L}(\overline{q}, \dot{\overline{q}})$. To justify the projection method we will try to show that it is possible to choose coordinates such that L_P and $\overline{L}$ coincide, at least up to a certain order. Then, we call the reduction a *natural reduction*. This topic will be considered in Section 6.5.

6.2   The abstract setting of Lagrangian systems

We consider a manifold $\mathcal{Q}$ modelled over a reflexive Banach space Q. Let $\mathcal{Q}_1$ be a manifold domain in $\mathcal{Q}$ and $\mathcal{Q}_2$ a manifold domain in $\mathcal{Q}_1$, modelled over the dense subspaces $Q_2 \subset Q_1 \subset Q$; forming again reflexive Banach spaces with norms $\| \cdot \|_1$ and $\| \cdot \|_2$ such that $\|q\| \leq c_1 \|q\|_1 \leq c_2 \|q\|_2$ for all $q \in Q_2$. Then let $\mathcal{Y}$ and $\mathcal{Y}^*$ be the following subsets of $T\mathcal{Q}$ and $T^*\mathcal{Q}$, respectively:

$$\mathcal{Y} = \bigcup_{q \in \mathcal{Q}_2} T_q \mathcal{Q}_1, \qquad \mathcal{Y}^* = \bigcup_{q \in \mathcal{Q}_2} T_q^* \mathcal{Q}_1.$$

$\mathcal{Y}$ is a manifold domain in $T\mathcal{Q}$ modelled over $Q_2 \times Q_1$. Here Q_1^* is the dual space of Q_1 in the Q_1–topology and not in the weaker Q–topology (e.g. $W^{s,p}(\Omega)^* = W^{s,q}(\Omega)$ with $1/p + 1/q = 1$). Thus, $Q_1^* \subset Q^*$ and $\mathcal{Y}^*$ can be understood as a manifold domain in $T^*\mathcal{Q}$ modelled over $Q_2 \times Q_1^*$. We call $\mathcal{Y}$ ($\mathcal{Y}^*$) the restriction of $T\mathcal{Q}$ ($T^*\mathcal{Q}$) to $(\mathcal{Q}_1, \mathcal{Q}_2)$.

A Lagrange function, or shortly a Lagrangian, on $\mathcal{Y}$ is a smooth function $L : \mathcal{Y} \to \mathbb{R}$. The Lagrangian system corresponding to L is obtained by considering the problem of making the functional

$$\mathcal{I}(q) = \int_{t_0}^{t_1} L(q(t), \dot{q}(t)) \, dt$$

stationary over the class of continuous functions $q : [t_0, t_1] \to \mathcal{Q}_2$ which additionally satisfy $q \in C^1([t_0, t_1], \mathcal{Q}_1)$, $q(t_0) = q_0$, and $q(t_1) = q_1$ for some fixed $q_0, q_1 \in \mathcal{Q}_2$. Whether q makes $\mathcal{I}$ stationary or not may be decided by inserting $\widetilde{q} = q + \varepsilon h + \mathcal{O}(\varepsilon^2) \in \mathcal{Q}_2$, where $h(t) \in T_{q(t)}\mathcal{Q}$ with $h(t_0) = h(t_1) = 0$, and taking the linear part in ε. Hence q makes $\mathcal{I}$ stationary if and only if

$$\delta \mathcal{I}(q)[h] = \int_{t_0}^{t_1} D_1 L(q, \dot{q})[h] + D_2 L(q, \dot{q})[\dot{h}] \, dt = 0$$

for all $h \in C^1([t_0, t_1], \mathcal{Y})$. Assuming we are in local coordinates, we can integrate the second term by parts which results in $0 = \int D_1 L(\cdots)[h] - \frac{d}{dt}(D_2 L(\cdots))[h] \, dt$. Since h is arbitrary we obtain the *Lagrange equations* corresponding to L:

$$\frac{d}{dt} D_2 L(q(t), \dot{q}(t)) = D_1 L(q(t), \dot{q}(t)).$$

To establish the connection to Hamiltonian systems we introduce the *fiber derivative* FL. Note that, for each $q \in \mathcal{Q}_2$, $L(q, \cdot)$ maps $T_q \mathcal{Q}_1$ into $\mathbb{R}$. Hence, we may define $FL : \mathcal{Y} \to \mathcal{Y}^*; (q, u) \to (q, D_2 L(q, u))$ where

$$D_2 L(q, u)[w] = \lim_{t \to 0} \frac{1}{t}(L(q, u + tw) - L(q, u)).$$

We call the Lagrangian L *regular close to a point* $y \in \mathcal{Y}$ if FL is a diffeomorphism when restricted to a suitable neighborhood $\mathcal{U}$ of y in $\mathcal{Y}$. Thus we have a local one–to–one correspondence between $(q, u) \in \mathcal{Y}$ and $(q, p) \in \mathcal{Y}^*$, where $p = D_2 L(q, u)$. Note that this is only possible if the model spaces $Q_2 \times Q_1$ and $Q_2 \times Q_1^*$ are isomorphic; in applications this will be guaranteed by choosing Q_1 as a Hilbert space.

Using the action $A : \mathcal{Y} \to I\!R$ and the energy $E : \mathcal{Y} \to I\!R$, given by

$$A(q, u) = D_2 L(q, u)[u], \qquad E = A - L,$$

we can introduce the Hamiltonian $H : \mathcal{Y}^* \subset T^*Q \to I\!R$ by $H(q, D_2 L(q, u)) = E(q, u)$. The mapping FL is the generalization of the *Legendre transformation* into the manifold context. Note also that the Hamilton function H admits a fiber derivative $FH : T^*Q \to TQ$, which, in the regular case, is just the inverse mapping of FL.

The classical examples for Lagrangian systems are the *simple mechanical systems*, where L is kinetic energy minus potential energy:

$$L(q, \dot{q}) = T(q, \dot{q}) - U(q) = \frac{1}{2}\langle \dot{q}, M(q)\dot{q}\rangle - U(q),$$

where M is the generalized mass matrix which defines a positive definite symmetric bilinear form on the configuration space Q (Riemannian metric). Then

$$p = M(q)\dot{q}, \quad A(q, \dot{q}) = \langle \dot{q}, M(q)\dot{q}\rangle, \quad \text{and } E = K + U.$$

Expressing E in terms of (q, p) yields $H(q, p) = \frac{1}{2}\langle p, M(q)^{-1}p\rangle + U(q)$, which is the associated Hamiltonian function.

Theorem 6.1
Assume that L is locally regular on $\mathcal{U} \subset \mathcal{Y} \subset TQ$. Let H be its associated Hamiltonian on $\mathcal{Y}^* \subset T^*Q$. Then a curve q with $(q, \dot{q}) \in \mathcal{U}$ satisfies the Lagrange equation of L if and only if $(q, p) = FL(q, \dot{q})$ satisfies the Hamilton equation corresponding to H and the canonical symplectic form ω_{can} on T^*Q.

The proof of this result is classical, when done in local coordinates (see [HM83, Ch.5.3]).

To obtain the vector field of the Lagrange equation on $\mathcal{Y} \subset TQ$ one can pull back the canonical symplectic form from T^*Q to $\mathcal{Y}$ using the fiber derivative FL. Then the Lagrange equation is the Hamiltonian system corresponding to the Hamiltonian $E : \mathcal{Y} \to I\!R$ and the symplectic form $\omega_L = FL_*\omega_{can}$.

6.3 Linear theory

We want to discuss the question under what conditions the flow on the center manifold of a Lagrangian system is again governed by a reduced Lagrangian problem. To shorten the formulations we introduce

Definition 6.2

The flow of a Hamiltonian system on a symplectic manifold $\mathcal{M}$ is called a Lagrangian flow, if there are canonical coordinates (q, p) on $\mathcal{M}$ such that the fiber derivative FH is regular. (Hence the Euler–Lagrange equation of the associated Lagrangian L generate the same flow.)

The question whether the flow on the center manifold is a Lagrangian flow is considerably more difficult than the corresponding one for Hamiltonian flows. As an example consider the Lagrangian $L(q, \dot{q}) = \frac{1}{2}(\dot{q}_1^2 - \dot{q}_2^2) + q_1 \dot{q}_2$ on $T\mathbb{R}^2$. It generates the Euler–Lagrange equations

$$\frac{d}{dt}(\dot{q}_1) = \dot{q}_2, \qquad \frac{d}{dt}(-\dot{q}_2 + q_1) = 0.$$

The eigenvalues of the associated linear operator are 0, 1, and -1 with eigenspaces

$$V(0) = \operatorname{span}\{(1, 0, 0, 0), (0, 1, 0, 0)\},$$

$$V(1) = \operatorname{span}\{(1, 1, 1, 1)\}, \quad V(-1) = \operatorname{span}\{(1, -1, -1, 1)\}.$$

On $X_1 = V(0)$ we find $\widetilde{H} \equiv 0$; and thus the reduced flow is the trivial one, which is not a Lagrangian flow.

Thus, we have obtained a first negative result:

In general, the flow on the center manifold of a Lagrangian system is not a Lagrangian flow.

The general quadratic Lagrangian L on $Q \times Q$ reads

$$L(q, \dot{q}) = \frac{1}{2}\langle M\dot{q}, \dot{q} \rangle + \langle Bq, \dot{q} \rangle + \frac{1}{2}\langle Cq, q \rangle,$$

where M and C can be taken as symmetric. Moreover, L is regular if and only if M is invertible. The Lagrange equation is given by

$$M\ddot{q} + (B - B^*)\dot{q} - Cq = 0. \tag{6.3}$$

This is the general form for linear Lagrange equations. For every equation of this form one finds a corresponding Lagrangian, which is not unique since the symmetric part of B does not enter in (6.3). We may also write this equation as a first order system:

$$\frac{d}{dt}\begin{pmatrix} q \\ \dot{q} \end{pmatrix} = K_L \begin{pmatrix} q \\ \dot{q} \end{pmatrix} = \begin{pmatrix} 0 & I \\ M^{-1}C & M^{-1}(B^* - B) \end{pmatrix} \begin{pmatrix} q \\ \dot{q} \end{pmatrix}.$$

The operator K_L on $Q \times Q$ corresponds to the linear operator JA defined in Chapter 3 (see also below); and we henceforth assume that it has a the same spectral properties, i.e.

there is a positive α such that in the strip $|\text{Re}\,\lambda| < \alpha$ the resolvent of K_L exists except for a finite number of points. Moreover the generalized eigenspaces of these eigenvalues are assumed to be finite–dimensional.

The fiber derivative relates $(q,\dot{q})$ to $(q,p) = (q, M\dot{q} + Bq)$. Hence, the action A and the energy E are given by $A(q,\dot{q}) = \langle M\dot{q} + Bq, \dot{q} \rangle$ and $E(q,\dot{q}) = A - L = \frac{1}{2}\langle M\dot{q}, \dot{q} \rangle - \frac{1}{2}\langle Cq, q \rangle$. Using $\dot{q} = M^{-1}(p - Bq)$ we obtain the Hamiltonian H on $Q \times Q^*$:

$$H(q,p) = \frac{1}{2}\langle M^{-1}p, p \rangle - \langle M^{-1}Bq, p \rangle + \frac{1}{2}\langle (B^*M^{-1}B - C)q, q \rangle.$$

The linear operator JA now satisfies

$$JA = \begin{pmatrix} M^{-1}B & M^{-1} \\ B^*M^{-1}B - C & -B^*M^{-1} \end{pmatrix} = \begin{pmatrix} I & 0 \\ B & M \end{pmatrix} K_L \begin{pmatrix} I & 0 \\ B & M \end{pmatrix}^{-1}, \quad J = \begin{pmatrix} 0 & I \\ -I & 0 \end{pmatrix}.$$

Asking for conditions guaranteeing that the center space has a Lagrangian flow we go back to the normal form theory provided in Section 3.2. Transferring the Lagrangian system into the associated Hamiltonian enables us to take full profit of the theory developed there. In particular, we know that the center space X_1 decomposes into a sum of invariant symplectic subspaces $W^k(is)$, $s \geq 0$, which are mutually orthogonal and contain at most four related Jordan chains, all of the same length. Hence, we first check whether each of the Hamiltonian systems in normal form, obtained by reduction onto $W^k(is)$, has a Lagrangian flow.

Lemma 6.3

All the normal forms of Section 3.2 generate a Lagrangian flow, except for the **Case 4** *with $m = 1$:*

Proof: In (q,p)–coordinates the operator K has the block structure $\begin{pmatrix} K_1 & K_2 \\ K_3 & K_4 \end{pmatrix}$. As $A = \Omega K = \begin{pmatrix} -K_3 & -K_4 \\ K_1 & K_2 \end{pmatrix}$ and $A^* = A$ we have $K_4 = -K_1^*$, $K_2^* = K_2$, and $K_3^* = K_3$.

For finding canonical coordinates $(\tilde{q}, \tilde{p})$ such that $\frac{\partial^2}{\partial p^2}\tilde{H}$ is invertible we use the transformation $\begin{pmatrix} q \\ p \end{pmatrix} = T\begin{pmatrix} \tilde{q} \\ \tilde{p} \end{pmatrix}$ with $T = \begin{pmatrix} G & F \\ D & E \end{pmatrix}$, which preserves the canonical symplectic structure (i.e. $TJT^* = J$) if and only if

$$GE^* - FD^* = I, \quad GF^* - FG^* = DE^* - ED^* = 0. \tag{6.4}$$

We will only use the case $G = E = I$ and $D = 0$, in which F can be any symmetric matrix. The new Hamiltonian $\tilde{H}$ is defined by $\tilde{A} = T^*AT$, and in particular the fourth block takes the form

$$\tilde{A}_4 = K_2 + K_1 F + F^* K_1^* - F^* K_3 F.$$

The flow generated by $K = JA$ is a Lagrangian flow, if and only if there is a T such that the (symmetric) matrix $\tilde{A}_4$ is invertible.

Taking $T = \begin{pmatrix} I & 0 \\ 0 & I \end{pmatrix}$ shows that the **Cases 1.1, 1.2,** and **3** generate Lagrangian flows. For **Case 2** let $T = \begin{pmatrix} I & F \\ 0 & I \end{pmatrix}$ with $F = \text{diag}\{N, \dots, N\}$ where $N = \begin{pmatrix} 0 & 0 \\ 0 & 1 \end{pmatrix}$. Then we obtain

$$\widetilde{A}_4 = \begin{pmatrix} \widehat{S} & N & & \\ N & \widehat{S} & \ddots & \\ & \ddots & \ddots & N \\ & & N & \widehat{S} \end{pmatrix}, \qquad \widehat{S} = \begin{pmatrix} 0 & s \\ s & 0 \end{pmatrix}.$$

As $\det \widetilde{A}_4 = (-s^2)^m$ and $s \neq 0$ this case is finished.

Case 4 with $m = 2$ can be handled by taking $F = I$. This leads to

$$\widetilde{A}_4 = \begin{pmatrix} 0 & 1 & & \\ 1 & 0 & 1 & \\ & 1 & 0 & 1 \\ & & 1 & 0 \end{pmatrix},$$

which is again invertible. For $m \geq 3$ we use $F = \widetilde{F}$ resulting in the invertible matrix $\widetilde{A}_4$:

$$\widetilde{F} = \begin{pmatrix} 1 & 0 & & & \\ 0 & 0 & 1 & & \\ & 1 & \ddots & \ddots & \\ & & \ddots & \ddots & 1 \\ & & & 1 & 0 \end{pmatrix}, \qquad \widetilde{A}_4 = \begin{pmatrix} 0 & 1 & & & \\ 1 & 0 & 0 & & \\ & 0 & 2 & \ddots & \\ & & \ddots & \ddots & 0 \\ & & & 0 & 2 \end{pmatrix}.$$

For $m = 1$ we have $K = 0$; thus this case does not generate a Lagrangian flow. Hence, the lemma is proved. $\qquad\square$

As an immediate consequence we obtain

Lemma 6.4
*If in the normal form of a linear Hamiltonian system the **Case 4** with $m = 1$ does not appear, then the flow on the center space is a Lagrangian flow.*

Proof: By the normal form theory of Chapter 3 the linear Hamiltonian system on the center space decomposes into decoupled systems on subspaces $W^k(is_k)$ which are pairwise orthogonal with respect to the symplectic structure. The previous result provides canonical coordinates (q_k, p_k) in each of these spaces such that the fiber derivative of the restriction of H onto $W^k(is_k)$ is regular. Putting together these results and using the decoupling Theorem 3.3 for the Hamiltonian we find that $(q, p) = (q_1, q_2, \dots, p_1, p_2, \dots)$ are canonical coordinates on the center space such that the fiber derivative is regular. $\square$

The converse of this lemma is not true, as the following example shows: Let $L(q, p) = \frac{1}{2}(\dot{q}_1^2 + \dot{q}_2^2) + q_1 \dot{q}_2$ be a Lagrangian on $T\mathbb{R}^2$. Then the corresponding eigenvalues are 0,

i, and $-i$. This shows that the full system is the center manifold, and hence it has a Lagrangian flow. However, as in the example above, the flow on $V(0)$ is the trivial one, which corresponds to Case 4 with $m = 1$. The normal form on $V(i) \oplus V(-i)$ is Case 1.2 with $m = 1$. Having this in mind we are now able to state the main result of this section:

Theorem 6.5

A linear Hamiltonian flow on a $2n$–dimensional center space X_1 is a Lagrangian flow if and only if the dimension of the kernel of the associated linear operator $K_1 : X_1 \to X_1$ is not larger than n, i.e.

$$dim \, kernel\,(K_1) \le \frac{1}{2}\,dim\,X_1. \tag{6.5}$$

Remarks: In fact, it can be shown that this result holds for arbitrary linear finite–dimensional Hamiltonian systems, not only under the assumption that all the eigenvalues are purely imaginary. However, since our normal form theory is restricted to center spaces (just for brevity) we content ourselves with the present result.

In [CM74, Ch.5.5] methods are given to transform any Hamiltonian system into a Lagrangian one. Yet, this is done by enlarging the system by introducing additional variables. Moreover, the resulting Lagrangian is not regular.

Proof: For a Lagrangian flow on $I\!\!R^{2n}$ the special form of the linear operator K_L shows that the kernel of K_L is given as $\{ (q,0) : Cq = 0 \}$; and hence $dim(kernel\,K_L) \le n$. This shows that the condition is necessary.

Using the normal form decomposition and the previous lemma we know that there are canonical coordinates $(q,p) = (q_1, q_2, p_1, p_2)$ such that H takes the form

$$H(q_1, q_2, p_1, p_2) = \frac{1}{2}\langle \begin{pmatrix} q_1 \\ q_2 \\ p_1 \\ p_2 \end{pmatrix}, \begin{pmatrix} 0 & 0 & 0 & 0 \\ 0 & A_1 & 0 & A_2 \\ 0 & 0 & 0 & 0 \\ 0 & A_2^* & 0 & A_4 \end{pmatrix} \begin{pmatrix} q_1 \\ q_2 \\ p_1 \\ p_2 \end{pmatrix} \rangle,$$

where A_4 is invertible. Here (q_2, p_2) spans the union of the good normal form cases while (q_1, p_1) corresponds to the union of all **Cases 4** with $m = 1$ (where H completely vanishes).

Assume first that $(q_1, p_1) \in I\!\!R^2$. For arbitrary $s = (s_2, \ldots, s_n)$ and $r = (r_2, \ldots, r_n)$ we define the matrix

$$T = \begin{pmatrix} 1 & -r^* & -r^*s & s^* \\ 0 & I & s & 0 \\ 0 & 0 & 1 & 0 \\ 0 & 0 & r & I \end{pmatrix}.$$

This matrix satisfies the relations in (6.4) and thus the mapping $(q,p) = T(\widehat{q},\widehat{p})$ is canonical. In these new coordinates the Hamiltonian $\widehat{H}$ is given by the symmetric matrix

$$\widehat{A} = T^*AT = \begin{pmatrix} 0 & 0 & 0 & 0 \\ 0 & A_1 & A_1 s + A_2 r & A_2 \\ 0 & s^*A_1 + r^*A_2^* & \delta & s^*A_2 + r^*A_4 \\ 0 & A_2^* & A_2^* s + A_4 r & A_4 \end{pmatrix}$$

where $\delta = s^*A_1 s + r^*A_2^* s + s^*A_2 r + r^*A_4 r$. Letting $r = A_4^{-1}A_2^* s$ we obtain

$$\widehat{A} = \begin{pmatrix} 0 & 0 & 0 & 0 \\ 0 & A_1 & (A_1 - A_2 A_4^{-1}A_2^*)s & A_2 \\ 0 & s^*(A_1 - A_2 A_4^{-1}A_2^*) & s^*(A_1 - A_2 A_4^{-1}A_2^*)s & 0 \\ 0 & A_2^* & 0 & A_4 \end{pmatrix}.$$

However, according to Lemma 6.4 the (q_2, p_2)–system is Lagrangian with the Lagrange function $L(q_2, \dot{q}_2) = \frac{1}{2}\langle A_4^{-1}\dot{q}_2, \dot{q}_2\rangle - \langle A_4^{-1}A_2 q_2, \dot{q}_2\rangle - \frac{1}{2}\langle (A_1 - A_2 A_4^{-1}A_2^*)q_2, q_2\rangle$. Since the total dimension of the kernel of K_L is not larger than n we necessarily have $(A_1 - A_2 A_4^{-1}A_2^*) \neq 0$. Thus, there is a vector $s \in \mathbb{R}^{n-1}$ such that $s^*(A_1 - A_2 A_4^{-1}A_2^*)s \neq 0$. But this implies that the fiber derivative of $\widehat{H}$ in the coordinates $(\widehat{q}, \widehat{p})$ is regular.

If the dimension of the (q_1, p_1)–space is larger, we can iteratively add each pair of zero eigenvalues in the same procedure as above, as long as $A_1 - A_2 A_4^{-1}A_2^*$ is different from zero. But this is true by the assumptions of the theorem. $\square$

Finally we want to remark an observation which is not yet understood properly. Note that the counterexample at the beginning of this section has an indefinite matrix M. So far the author was unable to find a Lagrangian system with positive definite matrix M not satisfying the dimension condition (6.5). However, this is exactly the important case for mechanical systems, where M represents a mass or density matrix, and also for elliptic problems, where the rank–one convexity implies the positive definiteness of M. For Lagrangian systems on $T\mathbb{R}^n$ with $n \leq 4$ and M positive definite it can be shown that (6.5) always holds. Thus, we state the following **conjecture**:

> Positive definiteness of M suffices to guarantee that the reduced flow on the center manifold is always Lagrangian.

The validity of this statement would be helpful for two reasons. One the one hand, almost all elliptic variational problems in mechanics are strongly elliptic which implies the positive definiteness of M. On the other hand, this would provide a natural class of Lagrangian systems which always have Lagrangian reduced problems. Only in such a class there is hope to find a direct (variational) connection between the full and the reduced problem, avoiding the use of Hamiltonian theory.

6.4   Nonlinear systems

We now consider the general nonlinear case. As in the Hamiltonian situation we always assume that the linear part has the properties specified in Section 6.3. Moreover, we require that the linearized flow on the center manifold is a Lagrangian flow, see (6.5). Then the following result holds.

Theorem 6.6
Let L be a Lagrangian on $\mathcal{Y} \subset TQ$ which is regular close to an equilibrium $q_0 \in Q$ of the Lagrange equation (i.e. $DL(q_0, 0) = 0$). Assume that the linearized flow on the center manifold is a Lagrangian flow. Then the (nonlinear) flow on the center manifold is also a Lagrangian flow.

Proof: By Theorem 4.1 we know that the flow on the center manifold is Hamiltonian as the associated Hamiltonian system satisfies all conditions necessary. Let $\widetilde{H}$ and $\widetilde{\omega}$ be the reduced Hamiltonian and the symplectic form on M_C, respectively. Since the linearized flow is a Lagrangian flow we know that there are coordinates (q, p) on M_C such that, in the equilibrium point $(q, p) = 0$, ω_0 has the canonical form and that $\frac{\partial^2}{\partial p^2} \widetilde{H}(0)$ is invertible. By Darboux's theorem 5.12 (with $G = \{e\}$) we know that (q, p) can even be chosen canonical in a whole neighborhood of 0. The invertibility of $\frac{\partial^2}{\partial p^2} \widetilde{H}$ also extends into a neighborhood; and hence the Legendre transform is locally possible. This yields the result. $\square$

Thus, we have established the

Center manifold reduction for Lagrangian systems:
We start with the original Lagrangian $L(q, \dot{q})$, where q lies in the large space Q. Using the Legendre transform $p = \partial L/\partial \dot{q}$ leads to the associated canonical Hamiltonian system with Hamiltonian $H(q, p)$. For the center manifold reduction we restrict H and the symplectic structure to $\widetilde{H}(x_1)$ and $\widetilde{\omega}_{x_1}$, $x \in M_C$, respectively. Using Darboux's theorem and Theorem 6.5 we find canonical coordinates $(\bar{q}_1, \bar{p}_1)$ on M_C, such that $\overline{H}(\bar{q}_1, \bar{p}_1)$ is locally regular. Thus, the inverse Legendre transform $\dot{\bar{q}}_1 = \partial \overline{H}/\partial \bar{p}_1$ yields the desired reduced Lagrangian $\overline{L}(\bar{q}_1, \dot{\bar{q}}_1)$.

We have summarized the method in the diagram in Figure 6.1.

6.5   Natural reduction

The center manifold reduction presented above picks out an exact solution manifold for the full problem. Yet, the reduced Lagrangian $\overline{L}(\bar{q}, \dot{\bar{q}})$ does not necessarily coincide with

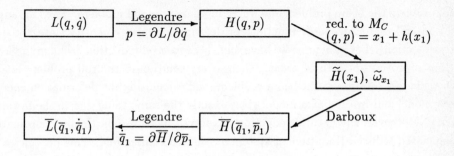

Figure 6.1: Reduction of Lagrangian systems

L_P of (6.2), even when the correct functions r and s, given by the center manifold, are used. Note, however, that the coordinates $(\overline{q}, \dot{\overline{q}})$ are not uniquely determined, since canonical changes of coordinates in the associated Hamiltonian system yield a whole family of possible coordinates. It seems likely that at least one choice of coordinates has the desired property $L_P = \widetilde{L}$. In this case we call the reduction a *natural reduction*. It is the major remaining challenge to solve the

Problem of natural reduction:

Assume that the flow on the center manifold M_C of a Lagrangian system is again Lagrangian, i.e. there exist coordinates $(\overline{q}, \dot{\overline{q}})$ in M_C and a reduced Lagrangian $\overline{L}$ on M_C such that the reduced flow is the flow generated by $\overline{L}$. Is it possible to choose the coordinates $(\overline{q}, \dot{\overline{q}})$ such that the projected Lagrangian $L_P = L_P(\overline{q}, \dot{\overline{q}})$, given by (6.2), coincides with $\overline{L}$?

One of the main difficulties appearing in this context is that the solutions on the center manifold cannot be characterized by a variational method directly. Even for the linear theory the question of natural reduction in general is still unsolved. Partial results are found throughout the applications in Part II.

The aim behind the idea of natural reduction is two–fold. On the one hand it provides a mathematical rigorous approach to projection methods. Particularly it gives the physically desirable property that the reduced problem has the same physical energy as the reduced one. On the other hand the natural reduction singles out a particular reduced Lagrangian, and thus avoids unnecessary ambiguity.

To make this more clear we turn our mind to the beam problem discussed in Chapter 11. There, the full Lagrangian problem corresponds to a quasilinear partial differential equation of nonlinear three–dimensional elastostatics, and the reduced problem to the twelve–dimensional ordinary differential equation governing the rod equations. Let us call the full problem the beam problem and the reduced one the rod problem. The Lagrangians are the so–called stored–energy functions or deformation energy density (integrated over the cross–section). Let us assume we have solved the natural reduction problem; in Section 11.4 this is proved up to third order. Then, every solution of the rod problem is also a corresponds to a solution of the beam problem, and calculating the deformation energy in the rod model and in the beam model gives exactly the same value due to the projection property. Moreover, the natural reduction has forced us to use particular coordinates in the associated reduced Hamiltonian system on the center manifold. This provides a help in defining the terms 'mean translation' and 'mean rotation' of a deformed cross–section.

In this context a complete mathematical theory can only be given for a weakened form of the problem, the *relaxed natural reduction problem*. There, we allow first to do a coordinate change in the full Lagrangian problem before posing the question of natural reduction.

Problem of relaxed natural reduction:

Assume that the flow on the center manifold M_C of a Lagrangian problem $L(q, \dot{q})$ is again Lagrangian. Is there a change of coordinates $(q, \dot{q}) = T(q, \dot{q})$ such that for the Lagrangian system $\underline{L}(q, \dot{q})$ the natural reduction is possible?

Recall that the Lagrangian $\underline{L}(q, \dot{q})$ is not equal to $L(T(q, \dot{q}))$, but has to be calculated via the Legendre transform and the associated canonical transformation in Hamiltonian formulation.

Using the results of Section 5.4 we obtain the following result.

Theorem 6.7
The relaxed natural reduction is always possible.

Proof: We start with the given Lagrangian $L(q, \dot{q})$ and the associated Hamiltonian $H(q, p)$. In Theorem 5.14 we have shown that there exist canonical coordinates $(\underline{q}, \underline{p}) = (\underline{q}_1, \underline{q}_2, \underline{p}_1, \underline{p}_2)$ such that the center manifold is given by $(\underline{q}_2, \underline{p}_2) \equiv 0$. Hence the reduced Hamiltonian is $\overline{H}(\underline{q}_1, \underline{p}_1) = \underline{H}(\underline{q}_1, 0, \underline{p}_1, 0)$. Doing, if necessary a further canonical transformation in $(\underline{q}_1, \underline{p}_1)$ and $(\underline{q}_2, \underline{p}_2)$ separately, we can assume that the fiber derivative is locally regular. The associated full Lagrangian $\underline{L}(q, \dot{q})$ is obtained from $\underline{H}(q, p)$ and the reduced Lagrangian $\overline{L}(\underline{q}_1, \dot{\underline{q}}_1)$ from $\overline{H}(\underline{q}_1, \underline{p}_1)$. However, on the center manifold M_C we find $\underline{L}(\underline{q}_1, 0, \dot{\underline{q}}_1, 0) = \langle \underline{p}_1, \dot{\underline{q}}_1 \rangle - \underline{H}(\underline{q}_1, 0, \underline{p}_1, 0)$ which is exactly the definition of $\overline{L}$. Thus, $\overline{L} = \underline{L}|_{M_C}$ which is the definition of natural reduction.   □

The construction of the coordinates $(\underline{q}_1, \underline{q}_2, \underline{p}_1, \underline{p}_2)$ is based on the full a–priori knowledge of the reduction function h. Hence, it is desireable to see how the approximate flattening of the center manifold, introduced in Section 4.3, can be used to give an approximation of the relaxed natural reduction method.

Theorem 6.8
Assume we have canonical coordinates $(q,p) = (q_1, q_2, p_1, p_2)$ such that the Hamiltonian satisfies $H(q,p) = H_1(q_1, p_1) + \mathcal{O}(\|(q_1, p_1)\|^m \|(q_2, p_2)\| + \|(q_2, p_2)\|^2)$. Let L_1 be the Legendre transform on H_1. Then, there are canonical coordinates $(\overline{q}_1, \overline{p}_1)$ on the center manifold M_C, such that the exact reduced Lagrangian $\overline{L}$ and the projected Lagrangian $L_P = L|_{M_C}$ satisfy

$$\overline{L}(\overline{q}_1, \dot{\overline{q}}_1) = L_1(\overline{q}_1, \dot{\overline{q}}_1) + \mathcal{O}(|(\overline{q}_1, \dot{\overline{q}}_1)|^{2m}),$$
$$L_P(q_1, \dot{q}_1) = L_1(q_1, \dot{q}_1) + \mathcal{O}(|(q_1, \dot{q}_1)|^{2m}).$$

Proof: The center manifold satisfies $(q_2, p_2) = h(q_1, p_1) = \mathcal{O}(|(q_1, p_1)|^m)$ and for the reduced Hamiltonian we obtain $\widetilde{H}(q_1, p_1) = H_1(q_1, p_1) + \mathcal{O}(|(q_1, p_1)|^{2m})$. Moreover, we can assume that the coordinates are already chosen such that the inverse Legendre transform is possible. From

$$\dot{q}_1 = \frac{\partial}{\partial p_1} H(q,p) = \frac{\partial}{\partial p_1} H_1(q_1, p_1) + \mathcal{O}(|(q_1, p_1)|^{m-1}\|(q_2, p_2)\| + \|(q_2, p_2)\|^2),$$
$$\dot{q}_2 = \frac{\partial}{\partial p_2} H(q,p) = \mathcal{O}(\|(q_2, p_2)\| + |(q_1, p_1)|^m),$$

we find that the Lagrangian has the form

$$L(q_1, q_2, \dot{q}_1, \dot{q}_2) = L_1(q_1, \dot{q}_1) + \mathcal{O}(|(q_1, \dot{q}_1)|^m |(q_2, \dot{q}_2)| + |(q_2, \dot{q}_2)|^2).$$

On M_C we have $(q_2, \dot{q}_2) = (r(q_1, \dot{q}_1), s(q_1, \dot{q}_1)) = \mathcal{O}(|(q_1, \dot{q}_1)|^m)$, and hence the projected Lagrangian L_P satisfies

$$L_P(q_1, \dot{q}_1) = L(q_1, r, \dot{q}_1, s) = L_1(q_1, \dot{q}_1) + \mathcal{O}(|(q_1, \dot{q}_1)|^{2m}).$$

To find an exact reduced Lagrangian we recall that $Dh(q_1, p_1) = \mathcal{O}(|(q_1, p_1)|^{m-1})$ implies the existence of canonical coordinates $(\overline{q}_1, \overline{p}_1) = (q_1, p_1) + \mathcal{O}(|(q_1, p_1)|^{2m-1})$ on the center manifold. The transformed Hamiltonian $\overline{H}$ satisfies $\overline{H}(\overline{q}_1, \overline{p}_1) = H_1(\overline{q}_1, \overline{p}_1) + \mathcal{O}(|(\overline{q}_1, \overline{p}_1)|^{2m})$. Doing the Legendre transformation results in $\overline{L}(\overline{q}_1, \dot{\overline{q}}_1) = L_1(\overline{q}_1, \dot{\overline{q}}_1) + \mathcal{O}(|(\overline{q}_1, \dot{\overline{q}}_1)|^{2m})$. $\square$

6.6 Symmetric Lagrangian systems

In addition we want to study systems being invariant under the action of a Lie group $\widetilde{G}$. As in the previous chapter we restrict ourself to the case that $\widetilde{G} = G \odot S$ acts via $\widetilde{\Phi}$

regular on $\mathcal{Q}$, such that the conditions (5.3) hold. According to the slice theorem 5.1 we are able to reduce the analysis to the case $\mathcal{Q} = G \times \widetilde{Q}$ where G acts as left translation on the G–component and S acts on $\widetilde{Q}$ via the linear representation τ_s, cf. (5.4). The induced action ϕ on $T\mathcal{Q}$ is given by

$$\phi_{g_1s}(g, \widetilde{q}, \dot{g}, \dot{\widetilde{q}}) = (g_1 g, \tau_s \widetilde{q}, DL_{g_1} \dot{g}, \tau_s \dot{\widetilde{q}}), \qquad g_1 s \in G \odot S = \widetilde{G}.$$

A Lagrangian L is $\widetilde{G}$–invariant if

$$L(g, \tau_s \widetilde{q}, \dot{g}, \tau_s \dot{\widetilde{q}}) = L(e, \widetilde{q}, DL_{g^{-1}} \dot{g}, \dot{\widetilde{q}}) \qquad \text{for all } gs \in G \odot S. \tag{6.6}$$

If, additionally, L is locally regular, then the fiber derivative FL satisfies

$$FL(g, \tau_s \widetilde{q}, \dot{g}, \tau_s \dot{\widetilde{q}})[(v, \tau_s w)] = FL(e, \widetilde{q}, DL_{g^{-1}} \dot{g}, \dot{\widetilde{q}})[(DL_{g^{-1}} v, w)]$$

and hence the action A and the energy E are also $\widetilde{G}$–invariant. Furthermore, the associated Hamiltonian H on $\mathcal{Y}^* \subset T^*\mathcal{Q}$ is invariant under the induced action.

As in the previous chapter we prefer to consider H as a function on $T^\circ(G \times \widetilde{Q}) = G \times \mathbf{g}^* \times \widetilde{Q} \times \widetilde{Q}^*$ rather than on $T^*(G \times \widetilde{Q})$. Similarly, the Lagrangian L can be considered as function on $T^\circ(G \times \widetilde{Q}) = G \times \mathbf{g} \times \widetilde{Q} \times \widetilde{Q}$ by identifying TG with $G \times \mathbf{g}$ by means of the mapping $DL_g : \mathbf{g} \to T_g G$. The coordinates in $T^\circ(G \times \widetilde{Q})$ and $T^\circ(G \times \widetilde{Q})$ are denoted by $(g, \xi, q, \dot{q})$ and (g, η, q, p), respectively. In the $\widetilde{G}$–invariant case we have $L = L(\xi, q, \dot{q})$ and $H = H(\eta, q, p)$; and the fiber derivative gives

$$\eta = \mathbf{d}_\xi L, \qquad p = \mathbf{d}_{\dot{q}} L, \qquad \xi = \mathbf{d}_\eta H, \qquad \dot{q} = \mathbf{d}_p H. \tag{6.7}$$

Hence, a $\widetilde{G}$–invariant Lagrangian system has an associated Hamiltonian system which is again G–invariant. According to Corollary 5.4 this property still holds for the reduced Hamiltonian system on the center manifold. Assuming that the flow on the center manifold is also a Lagrangian flow we know that there are some canonical coordinates, such that the inverse Legendre transformation $F\widetilde{H}$ leads to a reduced Lagrangian $\widetilde{L}$. However, the question is whether this change of coordinates can be chosen $\widetilde{G}$–invariant, as the one provided in Theorem 5.7. Only then the center manifold has a Lagrangian flow on a space $T(G \times Q_1)$ which is again $\widetilde{G}$–invariant.

Theorem 6.9
Let L be a $\widetilde{G}$–invariant locally regular Lagrangian on $T(G \times \widetilde{Q})$ with an equilibrium $x_0 = (e, \xi_0, 0, 0)$ admitting a finite–dimensional center manifold with a Lagrangian flow. Then the flow on the center manifold is diffeomorphic to the Lagrangian flow of a $\widetilde{G}$–invariant reduced Lagrangian $\widetilde{L}$ on $T(G \times Q_1)$ for some linear space Q_1.

Proof: Denote by $\widetilde{H}$ the reduced Hamiltonian and by $(q,p) \in Q_0 \times Q_0^*$ the canonical coordinates generating a Lagrangian flow on the center manifold by the locally regular fiber derivative $F\widetilde{H}$, i.e. the matrix $\frac{\partial^2}{\partial p^2}\widetilde{H}(0)$ is invertible. Of course there is a regular induced action ϕ of $\widetilde{G}$ on $Q_0 \times Q_0^*$. Denote the tangent space of the orbit $O_G(0)$ at the base point 0 by Q_G. The $\widetilde{G}$–invariance of $\widetilde{H}$ implies that the linearized Hamiltonian $\widehat{H}$ satisfies $\frac{\partial}{\partial q_G}\widehat{H} = 0$ for each $q_G \in Q_G$. Hence the intersection of the space Q_G with the space $\{0\} \times Q_0^*$ has to be trivial, otherwise $\frac{\partial^2}{\partial p^2}\widetilde{H}(0) = \frac{\partial^2}{\partial p^2}\widehat{H}(0)$ could not be invertible.

Since $(\{0\} \times Q_0^*)^\perp = \{0\} \times Q_0^*$ there is a symplectic basis $\{e_1, \ldots, e_{n+m}, \widetilde{e}_1, \ldots, \widetilde{e}_{n+m}\}$ of $Q_0 \times Q_0^*$, such that $\{e_1, \ldots, e_n\}$ and $\{\widetilde{e}_1, \ldots, \widetilde{e}_{n+m}\}$ are bases of Q_G and Q_0^*, respectively. Now we can identify Q_G with $\mathbf{g}$ and $\mathrm{span}\{e_{n+1}, \ldots, e_{n+m}\}$ with Q_1; then $Q_0 \times Q_0^*$ has the form $\mathbf{g} \times Q_1 \times \mathbf{g}^* \times Q_1^*$. As in the proof of Theorem 5.6 this choice generates a unique canonical $\widetilde{G}$–invariant symplectic form on $T^*(G \times Q_1)$. Furthermore, by the identity $Q_0^* = \mathbf{g}^* \times Q_1^*$ we know that the fiber derivative $F\widetilde{H}$ is locally regular. As $\widetilde{H}$ is still $\widetilde{G}$–invariant, so is the reduced Lagrangian $\widetilde{L}$ on $T(G \times Q_1)$. $\square$

6.7 Center manifolds for relative equilibria

In some physical applications involving $\widetilde{G}$–invariant Lagrangian or Hamiltonian systems one is interested in the flow in a neighborhood of a solution curve $g_0(t)$ in the Lie group G. If all the other components remain constant we call this a *relative equilibrium* (cf. [AM78, Def.4.3.6]). For an observer moving in G along the curve g_0 the system seems to be in equilibrium, and one can ask how the flow of the system behaves close to this point. In particular, the applications in elasticity in Chapters 10 and 11 are of this kind.

We consider the manifold $Q = G \times \widetilde{Q}$ and a G–invariant Hamiltonian H on $T^\circ(G \times \widetilde{Q}) = G \times \mathbf{g}^* \times \widetilde{Q} \times \widetilde{Q}^*$ equipped with its canonical symplectic form. Assume that $dH(e,0,0,0) = (0, \xi_0, 0, 0)$ with $\xi_0 \in \mathbf{g}$ and let g_0 be defined as solution of $\dot{g} = DL_g\xi_0$, $g(0) = e$; viz. $g_0(t) = \exp(t\xi_0)$. Then the curve $c_0(t) = (g_0(t), 0, 0, 0)$ is a solution of the Hamiltonian system. Of course, for any $h \in G$ the curve $g(t) = hg_0(t)$ is again a relative equilibrium. Thus it is no restriction to impose $g_0(0) = e$.

We want to study solutions (g, η, q, p) in $T^\circ(G \times \widetilde{Q})$ staying close to c_0. Therefore we first factor out the Lie group as in the previous chapter. According to (5.8) the full system is given by

$$\frac{d}{dt}\begin{pmatrix} g \\ \eta \\ q \\ p \end{pmatrix} = \begin{pmatrix} 0 & DL_g & 0 & 0 \\ -DL_g^* & j(\eta) & 0 & 0 \\ 0 & 0 & 0 & I \\ 0 & 0 & -I & 0 \end{pmatrix}\begin{pmatrix} 0 \\ \mathbf{d}_\eta H \\ \mathbf{d}_q H \\ \mathbf{d}_p H \end{pmatrix}. \tag{6.8}$$

Here $j(\eta) : \mathbf{g} \to \mathbf{g}^*$ is defined by $\langle j(\eta)\xi, \zeta \rangle_{\mathbf{g}} = \langle \eta, [\xi, \zeta] \rangle_{\mathbf{g}}$ for all $\xi, \zeta \in \mathbf{g}$. Hence, j generates the classical *Lie–Poisson bracket* on $\mathbf{g}^*$ ([Ma81, Ar78]). Note that j is linear in $\eta \in \mathbf{g}^*$.

The equation for (η, q, p) decouples and forms an independent generalized Hamiltonian system. Moreover, $(\eta, q, p) = (0, 0, 0)$ is an equilibrium having the linearization

$$\dot{\eta} = j(\eta)\xi_0 , \qquad \frac{d}{dt}\begin{pmatrix} q \\ p \end{pmatrix} = \begin{pmatrix} 0 & I \\ -I & 0 \end{pmatrix}\begin{pmatrix} A_{qq} & A_{qp} \\ A_{qp}^* & A_{pp} \end{pmatrix}\begin{pmatrix} q \\ p \end{pmatrix}, \qquad (6.9)$$

where the Hamiltonian has the form

$$H(\eta, q, p) = \langle \eta, \xi_0 \rangle + \frac{1}{2}\langle \begin{pmatrix} \eta \\ q \\ p \end{pmatrix}, \begin{pmatrix} A_{\eta\eta} & A_{\eta q} & A_{\eta p} \\ A_{\eta q}^* & A_{qq} & A_{qp} \\ A_{\eta p}^* & A_{qp}^* & A_{pp} \end{pmatrix}\begin{pmatrix} \eta \\ q \\ p \end{pmatrix} \rangle + \mathcal{O}(\|(\eta, q, p)\|^3). \quad (6.10)$$

Thus, the linearized system (6.9) has decoupled further. A closely related method called *block diagonalization* was developed in [MSLP89] to study the stability properties of relative equilibria.

The center space of (6.9) is the sum of a subspace X_0 of $\mathbf{g}^*$ and a subspace X_1 of $\widetilde{Q} \times \widetilde{Q}^*$. According to Chapter 3, $\widetilde{Q} \times \widetilde{Q}^*$ decomposes into $X_1 \times X_2$ with X_1 and X_2 being both symplectic with respect to the canonical symplectic structure. X_0 is the center space of the linear operator

$$C_{\xi_0} : \begin{cases} \mathbf{g}^* & \to & \mathbf{g}^*, \\ \eta & \to & j(\eta)\xi_0 . \end{cases}$$

In [AM78, MSLP89] this operator is called the *infinitesimal adjoint action* and denoted by $ad_{\xi_0}\eta$.

For reasons becoming clear later we additionally assume $X_0 = \mathbf{g}^*$, i.e.

$$\text{all the eigenvalues of } C_{\xi_0} \text{ have zero real part.} \qquad (6.11)$$

Generally this is not the case as the first of the following examples shows. (However, restricting G to an appropriate subgroup $\widehat{G}$ containing $g_0(t)$ leads to a $\widehat{G}$–invariant problem such that the corresponding operator $\widehat{C}_{\xi_0}$ on the Lie algebra $\widehat{\mathbf{g}}^*$ of $\widehat{G}$ has the desired property.) The second example is a preparation for Chapter 11.

Example 1: Let $G = \mathbb{R}^+ \times \mathbb{R}$ with product $(a, b) \cdot (c, d) = (ac, ad + b)$. Then the Lie algebra $\mathbf{g}$ is given by $\mathbb{R}^2$ with the Lie bracket $[(\alpha, \beta), (\gamma, \delta)] = (0, \alpha\delta - \beta\gamma)$. With $\xi = (\xi_1, \xi_2)$ we obtain, by using $\langle C_\xi \eta, \zeta \rangle = \langle \eta, [\xi, \zeta] \rangle$, the representation

$$C_\xi \begin{pmatrix} \alpha \\ \beta \end{pmatrix} = \begin{pmatrix} 0 & -\xi_2 \\ 0 & \xi_1 \end{pmatrix}\begin{pmatrix} \alpha \\ \beta \end{pmatrix}$$

for all $(\alpha, \beta) \in \mathbb{R}^2$. Hence, condition (6.11) holds if and only if $\xi_1 = 0$.

Example 2: We consider the Lie group of Euclidian transformations on $\mathbb{R}^3$, $SO(3) \times \mathbb{R}^3$. The Lie algebra is

$$so(3) \times \mathbb{R}^3 = \{ (A, a) \in L(\mathbb{R}^3, \mathbb{R}^3) \times \mathbb{R}^3 \; : \; A + A^* = 0 \}$$

with the Lie bracket $[(A, a), (B, b)] = (AB - BA, Ab - Ba)$. Introducing the coordinates $c = (c_1, \ldots, c_6)$ by $(A, a) = (\begin{pmatrix} 0 & c_3 & -c_2 \\ -c_3 & 0 & c_1 \\ c_2 & -c_1 & 0 \end{pmatrix}, \begin{pmatrix} c_4 \\ c_5 \\ c_6 \end{pmatrix})$ the Lie bracket transforms into

$$[\begin{pmatrix} c_1 \\ \vdots \\ c_6 \end{pmatrix}, \begin{pmatrix} d_1 \\ \vdots \\ d_6 \end{pmatrix}] = \begin{pmatrix} c_3 d_2 - c_2 d_3 \\ c_1 d_3 - c_3 d_1 \\ c_2 d_1 - c_1 d_2 \\ c_3 d_5 - c_2 d_6 - c_5 d_3 + c_6 d_2 \\ c_1 d_6 - c_3 d_4 - c_6 d_1 + c_4 d_3 \\ c_2 d_4 - c_1 d_5 - c_4 d_2 + c_5 d_1 \end{pmatrix}.$$

Hence, for a fixed $\xi = (\xi_1, \ldots, \xi_6)$ the operator C_ξ takes the form

$$C_\xi \begin{pmatrix} \eta_1 \\ \vdots \\ \eta_6 \end{pmatrix} = \begin{pmatrix} 0 & -\xi_3 & \xi_2 & 0 & -\xi_6 & \xi_5 \\ \xi_3 & 0 & -\xi_1 & \xi_6 & 0 & -\xi_4 \\ -\xi_2 & \xi_1 & 0 & -\xi_5 & \xi_4 & 0 \\ 0 & 0 & 0 & 0 & -\xi_3 & \xi_2 \\ 0 & 0 & 0 & \xi_3 & 0 & -\xi_1 \\ 0 & 0 & 0 & -\xi_2 & \xi_1 & 0 \end{pmatrix} \begin{pmatrix} \eta_1 \\ \vdots \\ \eta_6 \end{pmatrix}.$$

This shows that C_ξ has exactly the eigenvalues 0 and $\pm i |(\xi_1, \xi_2, \xi_3)|$, each of them being double. Thus the condition (6.11) holds for all ξ. The application to Saint–Venant's problem in Chapter 11 involves the vector $\xi = (0, 0, 0, 0, 0, 1)$ which implies that C_ξ has only the eigenvalue 0.

If the condition (6.11) is satisfied and if the nonlinear (η, q, p)–system of (6.8) has a center manifold, then it has to be the graph of a function $h : U \to X_2$ where U is a neighborhood of zero in $\mathbf{g}^* \times X_1$. By adding the g–component we find that

$$M_C = \{ (g, \eta, x_1 + h(\eta, x_1)) \in T^\circ(G \times \widetilde{Q}) \; : \; (g, \eta, x_1) \in G \times U \}$$

is an invariant manifold for the full system (6.8). We call M_C the center manifold of the relative equilibrium $(\eta, q, p) = 0$. Moreover, since the tangent space of M_C at $(e, 0, 0, 0)$ is given by $\mathbf{g} \times \mathbf{g}^* \times X_1$ with $X_1 \subset \widetilde{Q} \times \widetilde{Q}^*$ being symplectic, we immediately see that the restriction of ω_{can} onto TM_C is nondegenerate. Hence, Theorem 4.1 is applicable and we obtain a reduced Hamiltonian system, which is automatically G–invariant by construction.

Here we see that it is important to have all eigenvalues of C_{ξ_0} on the imaginary axis which is equivalent to $X_0 = \mathbf{g}^*$. Otherwise the center manifold can not be symplectic, unless G is restricted to a subgroup $\widehat{G}$ such that $\widehat{\mathbf{g}} \times X_0$ is symplectic.

Additionally, we can apply Theorem 5.7 to construct a local diffeomorhism between M_C and $T^\circ(G \times Q_1)$, where Q_1 is any Banach space with $2 \dim Q_1 = \dim X_1$, such that the restricted symplectic form on M_C is transformed into the canonical one.

Now we turn to Lagrangian systems and study the question whether a Lagrangian system can be reduced onto a center manifold in the neighborhood of a relative equilibria $g_0(t)$. To this end consider the $\widetilde{G}$–invariant Lagrangian $L = L(\xi, q, \dot{q})$ on $T^\circ(G \times \widetilde{Q})$ such that $\mathbf{d}L$ vanishes in $(\xi_0, 0, 0)$. The fiber derivative FL is regular close to $(\xi_0, 0, 0)$ if and only if the second derivative $D^2_{(\xi, \dot{q})}L$ is invertible in $(\xi_0, 0, 0)$. Then the associated Hamiltonian on $T^\circ(G \times \widetilde{Q})$ satisfies $\mathbf{d}H(0, 0, 0) = (\xi_0, 0, 0)$ and we are in the situation described above. Hence, under the assumption (6.11) for C_{ξ_0} the reduction process can be performed to obtain a reduced G–invariant Hamiltonian system on the center manifold equipped with canonical coordinates and the canonical symplectic structure.

If it is possible to show that the reduced Hamiltonian $\widetilde{H} = \widetilde{H}(\eta, q_1, p_1)$ has a locally regular fiber derivative, i.e. $A_4 = D^2_{(\eta, p_1)}\widetilde{H}(0)$ is invertible, then we know that the flow on the center manifold is also a Lagrangian flow. This condition is of course necessary and sufficient. Yet, in practice it might be very hard to calculate the second derivative A_4. In the applications given below, only the first example in Chapter 8 allows a simple calculation of A_4. Already the necking problem in Chapter 10 involves lengthy computations to obtain A_4. For that reason we develop another condition which is completely based on spectral properties of an associated linear operator. This gives a considerable simplification as the necessary informations are usually derived much earlier in the process of calculating the center space.

To obtain a related linear probem we use a coordinate system moving on G along g_0. Therefore we define the new variable

$$h(t) = g_0^{-1}(t)g(t). \tag{6.12}$$

Let R_g be the right translation on G, i.e. $R_h g = gh$, then the product rule gives $(gh)\dot{} = DR_h \dot{g} + DL_g \dot{h}$ and $(g^{-1})\dot{} = -DR_{g^{-1}} DL_{g^{-1}} \dot{g}$. Hence, for $h(t) = g_0^{-1}(t)g(t)$ we obtain the relation

$$\dot{h} = DR_g(g_0^{-1})\dot{} + DL_{g_0^{-1}} \dot{g} = -DR_g DR_{g_0^{-1}} DL_{g_0^{-1}} \dot{g}_0 + DL_{g_0^{-1}} \dot{g}$$

$$= -DR_{g_0^{-1} g}\xi_0 + DL_{g_0^{-1} g} DL_{g^{-1}} \dot{g} = DL_h(\mathbf{d}_\eta H(\eta, q, p) - DL_{h^{-1}} DR_h \xi_0).$$

The expression $DL_{h^{-1}} DR_h \xi_0 = Ad_{h^{-1}}\xi_0$ is equal to $\mathbf{d}_\eta N_0$ when $N_0 = N_0(g, \eta)$ is defined by $\langle \eta, Ad_{g^{-1}}\xi_0 \rangle = \langle Ad_g^*\eta, \xi_0 \rangle$. Note that L_g always denotes the left translation whereas $L = L(\xi, q, \dot{q})$ denotes the Lagrangian.

Proposition 6.10

a) *The function N_0 is constant along all solutions of any G–invariant Hamiltonian system on $T^\circ(G \times \widetilde{Q})$.*

b) *Consider N_0 as a Hamiltonian function on $T^\circ(G \times \widetilde{Q})$, then its flow is just the left translation by $g_0(t) = \exp(t\xi_0)$.*

c) *For $g = \exp(s\xi) = e + s\xi + \mathcal{O}(s^2)$ we have the expansion*

$$N_0(g(s), \eta) = \langle \eta_0, \xi \rangle + s\langle \eta, [\xi_0, \xi] \rangle + \mathcal{O}(s^2) = \langle \eta_0, \xi \rangle + s\langle C_{\xi_0}\eta, \xi \rangle + \mathcal{O}(s^2).$$

Proof: Since assertion a) is not needed in the following we desist from giving a proof and refer to [AM78, Thm.4.4.3].

For b) we easily obtain $\dot{g} = DL_g \mathbf{d}_\eta N_0 = DR_g \xi_0$ which has the general solution $g(t) = \exp(t\xi_0)g(0)$. It remains to be shown that $\dot{\eta} = -DL_g^* \mathbf{d}_g N_0 + j(\eta)\mathbf{d}_\eta N_0 = 0$ for all (g, η). Testing with $\xi \in \mathbf{g}$ we have to satisfy

$$
\begin{aligned}
0 &= \langle \dot{\eta}, \xi \rangle = \langle j(\eta)\mathbf{d}_\eta N_0, \xi \rangle + \langle -DL_g^* \mathbf{d}_g N_0, \xi \rangle \\
&= \langle \eta, [\mathbf{d}_\eta N_0, \xi] \rangle - \langle \mathbf{d}_g N_0, DL_g \xi \rangle \\
&= \langle \eta, [Ad_{g^{-1}}\xi_0, \xi] \rangle - \langle \eta, D_g (Ad_{g^{-1}}\xi_0) [DL_g \xi] \rangle.
\end{aligned}
$$

Using the formula of the lemma given below the proof of b) is completed.

The same formula, evaluated at $g = e$, results in $\mathbf{d}_g N_0(e, \eta)[\xi] = \langle \eta, [\xi_0, \xi] \rangle$; and the expansion in c) follows. □

Lemma 6.11

For all $\xi, \zeta \in \mathbf{g}$ and all $g \in G$ the relation $D_g (Ad_{g^{-1}}\zeta)[DL_g \xi] = [Ad_{g^{-1}}\zeta, \xi]$ holds.

Proof: We use the exponential function $\exp(t\xi)$ and the action $\varphi_g(h) = ghg^{-1}$. Then,

$$
\begin{aligned}
D_g (Ad_{g^{-1}}\zeta)[DL_g \xi] &= \tfrac{d}{ds} \left\{ Ad_{(g\exp(s\xi))^{-1}}\zeta \right\}_{s=0} \\
&= \tfrac{d}{ds} \left\{ Ad_{\exp(-s\xi)} (Ad_{g^{-1}}\zeta) \right\}_{s=0} = D_g \left(Ad_{g^{-1}}\widehat{\zeta} \right)[\xi],
\end{aligned}
$$

where $\widehat{\zeta} = Ad_{g^{-1}}\zeta$. Thus it is sufficient to prove the formula in the case for $g = e$.

Letting $g(s, t) = \exp(-s\xi) \exp(t\zeta) \exp(s\xi) \exp(-t\zeta)$ we have $\tfrac{d}{ds}g(s, t)|_{s,t=0} = 0$ and, according to Chapter 5, $\tfrac{d^2}{ds\,dt}g(s, t)|_{s,t=0} = -[\xi, \zeta]$. This yields

$$
\begin{aligned}
D_g (Ad_{g^{-1}}\zeta) [\xi]|_{g=e} \\
= \tfrac{d}{dt} \left\{ \tfrac{d}{ds} (\exp(-s\xi)\exp(t\zeta)\exp(s\xi))_{s=0} \right\}_{t=0} &= \tfrac{d^2}{ds\,dt} \left\{ g(s, t)\exp(t\zeta) \right\}_{s,t=0} \\
= \left\{ DR_{\exp(t\zeta)} \tfrac{d^2}{ds\,dt}g(s, t) \right\}_{s,t=0} &+ \tfrac{d}{dt} \left(DR_{\exp(t\zeta)} \tfrac{d}{ds}g(s, t)|_{s=0} \right)_{t=0} \\
= -[\xi, \zeta] = [\zeta, \xi] &= [Ad_{g^{-1}}\zeta, \xi]|_{g=e}.
\end{aligned}
$$

This proves the lemma.   □

Since the flow generated by N_0 is just a left translation it commutes with the flow of every $\widetilde{G}$–invariant Hamiltonian. Moreover, substracting N_0 from H we obtain the *augmented Hamiltonian*

$$H_{aug}(g,\eta,q,p) = H(\eta,q,p) - N_0(g,\eta).\qquad (6.13)$$

The augmented Hamiltonian H_{aug} generates a flow which is the composition of the flow defined by H and the left translation by $g_0^{-1}(t) = \exp(-t\xi_0)$. This means that the Hamiltonian systems generated by $H(\eta,q,p)$ and by $H_{aug}(h,\eta,q,p)$ are equivalent in the sense that a solution of one system transfers via (6.12) into a solution of the other system.

The advantage of considering the augmented Hamiltonian system is that the relative equilibrium g_0 corresponds, in the moving system, to the real equilibrium $(e,0,0,0)$. This is equivalent to saying that $\mathbf{d}H(e,0,0,0) = (0,\xi_0,0,0) \neq 0$ and $\mathbf{d}H_{aug}(e,0,0,0) = 0$. Thus the linear theory developed above is applicable to the augmented system only. However the transformation into the moving coordinates destroys the $\widetilde{G}$–invariance of the system. N_0 really depends on h unless G is an Abelian group, where $Ad_g\xi = \xi$. Thus it is necessary to consider both systems simultaneously in order to keep track of the $\widetilde{G}$–invariance in the nonmoving system, but to study the linearization in the augmented system.

With (6.10) and Prop. 6.10 $H_{aug}(\exp(s\xi), s\eta, sq, sp)$ has the expansion

$$H_{aug} = \frac{s^2}{2}\langle \begin{pmatrix} \xi \\ \eta \\ q \\ p \end{pmatrix}, \begin{pmatrix} 0 & -C_{\xi_0} & 0 & 0 \\ -C_{\xi_0}^* & A_{\eta\eta} & A_{\eta q} & A_{\eta p} \\ 0 & A_{\eta q}^* & A_{qq} & A_{qp} \\ 0 & A_{\eta p}^* & A_{qp}^* & A_{pp} \end{pmatrix} \begin{pmatrix} \xi \\ \eta \\ q \\ p \end{pmatrix} \rangle + \mathcal{O}(|s|^3).\qquad (6.14)$$

Since the fiber derivative is locally regular if and only if $D^2_{(\eta,p)}H_{aug}(e,0)$ is invertible we conclude that H_{aug} is locally regular if and only if H is so. We remark that the linearized problem (6.9) of the nonmoving systems does not depend on $A_{\eta\eta}$. Thus, it is not suitable for answering the question of invertibility of the fiber derivative.

Theorem 6.12
Let L be a $\widetilde{G}$–invariant Lagrangian on $T^\circ(G \times \widetilde{Q})$ such that $\mathbf{d}L(\xi_0,0,0) = 0$ for some $\xi_0 \in \mathbf{g}$ with C_{ξ_0} satisfying condition (6.11) and such that L is regular in a neighborhood of $(\xi_0,0,0)$. Moreover, assume that there exists a finite–dimensional center manifold M_C for the associated Hamiltonian system (6.8). If, additionally, the flow on the center space of the linearized augmented Hamiltonian system is a Lagrangian flow, then the reduced flow on the center manifold is a Lagrangian flow generated by a $\widetilde{G}$–invariant reduced Lagrangian $\widetilde{L}$.

Remark: This theorem together with the kernel condition (6.5) of Theorem 6.5 for the augmented system enables us to prove the existence of a reduced Lagrangian on the center manifold, just by deriving the spectral properties of the linearized augmented system.

Proof: We consider the flow of H on $T^\circ(G \times \widetilde{Q})$ and of the reduced Hamiltonian $\widetilde{H}$ on the center manifold $M_C = T^\circ(G \times Q_1)$. Augmenting both systems by subtracting N_0 we obtain the Hamiltonians H_{aug} and $\widetilde{H}_{aug}$. However, in both cases the flows are related by $g(t) = g_0(t)h(t)$, hence the flow generated by $\widetilde{H}_{aug}$ is just the center manifold flow corresponding to the flow of H_{aug}.

Now, the assumption, that the center space of H_{aug} has a Lagrangian flow, implies that $\widetilde{H}_{aug}$ is locally regular. This shows that the fiber derivative of $\widetilde{H}$ is also regular in a neighborhood of zero. The $\widetilde{G}$–invariance of $\widetilde{H}$, and hence of $\widetilde{L}$, is a consequence of Theorem 6.9.   □

Chapter 7

Nonautonomous systems

7.1 First considerations

Time–dependent Hamiltonian systems are defined completely analogous to autonomous ones, with the only difference that the Hamiltonian function H may depend on the time variable t, viz.

$$\dot{x} = J(x)\mathbf{d}_x H(x, t) \tag{7.1}$$

In these systems the Hamiltonian is no longer a conserved quantity; as is seen by

$$\frac{d}{dt}H(x(t), t) = \frac{\partial}{\partial t}H(x(t), t) + \langle \mathbf{d}_x H, J(x)\mathbf{d}_x H(x, t)\rangle = \frac{\partial}{\partial t}H(x(t), t).$$

The motivation for studying these systems stems from nonautonomous Lagrangian systems $L = L(q, \dot{q}, t)$ on TQ. Making the integral $I(q) = \int_r^s L(q(t), \dot{q}(t), t)dt$ stationary yields the condition $\frac{d}{dt}\left(\frac{\partial L}{\partial \dot{q}}\right) - \frac{\partial L}{\partial q} = 0$, as in the autonomous case. Defining the conjugate momentum $p = \frac{\partial L}{\partial \dot{q}}$ and the Hamiltonian by $H(q, p, t) = \langle \dot{q}, p\rangle - L(q, \dot{q}, t)$ we easily find the classical Hamiltonian equations in the form

$$\dot{q} = \frac{\partial H}{\partial p}(q, p, t), \ \dot{p} = -\frac{\partial H}{\partial q}(q, p, t).$$

A more general treatment of nonautonomous systems, using contact forms, is given in [Ar78].

For simplicity we do not consider any symmetries in this chapter. There would be no difficulties in combining the present theory with the one in the previous two chapters, but we want to keep the presentation more comprehensive. Hence, we may assume that the phase space is the Banach space X and that J is generated by a symplectic structure ω on X. Again we consider the situation that the system has the equilibrium solution $x = 0$. Moreover we assume that the linearized problem $\dot{x} = J(0)Ax$ is time–independent. Then under appropriate uniformity assumptions on the t–dependence (see [Mi86a]) a center manifold of the form

$$M_C = \{ (x,t) = (x_1, h(x_1,t),t) \in \mathbb{R} \times X \; : \; \|x_1\| < \varepsilon, \, t \in \mathbb{R} \}$$

is known to exist. Note that M_C is defined as graph over the infinite cylinder generated by the product of a neighborbood $U_1 \subset X_1$ of zero and the time–axis $\mathbb{R}$. We have $h(x_1,t) = \mathcal{O}(\|x_1\|^2)$ and $Dh(x_1,t) = \mathcal{O}(\|x_1\|)$ for $x_1 \to 0$, uniformly in $t \in \mathbb{R}$. Hence, according to Theorem 4.1 the restriction of the symplectic form ω to $M_C \cap \{t = t_0\}$ is locally nondegenerate for all t_0. Thus we may write the reduced system in the form

$$\dot{x}_1 = \tilde{J}_1(x_1,t)\mathrm{d}_{x_1}\tilde{H}(x_1,t). \tag{7.2}$$

Now the question is whether suitable coordinates on M_C exist such that (7.2) is again a Hamiltonian system of the form (7.1). This means we have to find a t-dependent transformation which makes $\tilde{J}$ time–independent. Then the results of the previous chapters provide even canonical coordinates and enables us to transforms the reduced system back into a Lagrangian one.

7.2   The extended system and Darboux's theorem

To find the desired coordinates we use the classical method of extending the phase space by introducing the additional variables e and t. On $X \times \mathbb{R}^2$ we define the extended symplectic structure

$$\omega^{ext}_{(x,e,t)}((u_1,v_1,w_1),(u_2,v_2,w_2)) = \omega_x(u_1,u_2) + v_1 w_2 - w_1 v_2$$

and the extended Hamiltonian

$$H^{ext}(x,e,t) = H(x,t) - e.$$

Obviously the extended vector field is

$$\frac{d}{dt}\begin{pmatrix} x \\ e \\ t \end{pmatrix} = \begin{pmatrix} J(x) & 0 & 0 \\ 0 & 0 & 1 \\ 0 & -1 & 0 \end{pmatrix}\mathrm{d}_{(x,e,t)}H^{ext} = \begin{pmatrix} J(x)\mathrm{d}_x H(x,t) \\ \mathrm{d}_t H(x,t) \\ 1 \end{pmatrix}.$$

Now the reduced problem on the center manifold may be extended in the same way: $(x,e,t) = (x_1, h(x_1,t), e, t)$ with the extended reduced Hamiltonian $\tilde{H}^{ext}(x_1,e,t) = \tilde{H}(x_1,t) - e$. Writing the operator Ω^{ext} associated to ω^{ext} with respect to (x_1,x_2,e,t) we have

$$\Omega^{ext}(x,e,t) = \begin{pmatrix} \Omega_1(x) & \Omega_{12}(x) & 0 & 0 \\ -\Omega_{12}^*(x) & \Omega_2(x) & 0 & 0 \\ 0 & 0 & 0 & 1 \\ 0 & 0 & -1 & 0 \end{pmatrix};$$

and the reduced symplectic structure takes the form

$$\widetilde{\Omega}^{ext} = \begin{pmatrix} \widetilde{\Omega}_1(x_1,t) & 0 & r(x_1,t) \\ 0 & 0 & 1 \\ -r^*(x_1,t) & -1 & 0 \end{pmatrix} \tag{7.3}$$

with $\widetilde{\Omega}_1 = \Omega_1 - D_{x_1}h^*\Omega_{12}^* + \Omega_{12}D_{x_1}h + D_{x_1}h^*\Omega_2 D_{x_1}h$ and $r = (\Omega_{12} + D_{x_1}h^*\Omega_2)D_t h$, where h and Ω are evaluated at (x_1,t) and $(x_1, h(x_1,t))$, respectively. Of course, $\widetilde{\Omega}^{ext}$ is nondegenerate on the whole infinite cylinder.

Now we show that the symplectic structure $\widetilde{\omega}^{ext}$ can be changed into the canonical one by a change of coordinates of the form $(x_1, e, t) \rightarrow (\phi(x_1,t), e + \psi(x_1,t), t)$. Here the time variable is not changed, i.e. the old and the new time coincide. Furthermore the transformation will be calculated locally in time; this is important in order to maintain the qualitative t–behavior of the system. This provides a generalization of Darboux's theorem to nonautonomous symplectic structures.

Theorem 7.1

There is a transformation $\Phi : (x_1, e, t) \rightarrow (\phi(x_1,t), e + \psi(x_1,t), t)$ *being defined on a cylinder* $\{\|x_1\| < \varepsilon\} \times I\!\!R^2$ *such that in the new coordinates* $(y, f, t) = \Phi(x_1, e, t)$ *the symplectic form* $\widetilde{\omega}^{ext}$ *is the canonical one (i.e* $\Phi_*\omega_{can} = \widetilde{\omega}^{ext}$). *Moreover for each* $t_0 \in I\!\!R$, *the mapping* $\Phi(\cdot, \cdot, t_0)$ *is determined only by* $\widetilde{\Omega}^{ext}(\cdot, t_0)$ *and by* $\frac{d}{dt}\widetilde{\Omega}^{ext}(\cdot, t_0)$.

Proof: Obviously for each t_0 $\widetilde{\Omega}_1(\cdot, t_0)$ in (7.3) is locally a symplectic form over the x_1–space. Hence using the construction in Darboux's theorem we find a transformation $x_1 \rightarrow y = \phi(x_1,t)$ sucht that $D_{x_1}\phi^*\widetilde{\Omega}_1 D_{x_1}\phi \equiv \Omega_{1\,can}$. Hence the mapping $(y, e, t) = \widehat{\Phi}(x_1, e, t) = (\phi(x_1,t), e, t)$ transforms $\widetilde{\Omega}^{ext}(x_1, e, t)$ into $\widehat{\Omega}(y, e, t)$ having still the same structure as given in (7.3) but now with $\widehat{\Omega}_1 = \Omega_{1\,can}$ and $\widehat{r}(y,t) = r + (D_{x_1}\phi)^*\widetilde{\Omega}_1 \frac{d}{dt}\phi$ instead of $\widetilde{\Omega}_1$ and r, respectively.

Since $\widehat{\Omega}_1$ is constant the matrix $\widehat{\Omega}$ defines a closed two–form if and only if $D_y\widehat{r} - (D_y\widehat{r})^* = 0$. However this implies that $\widehat{r}(y,t) = -D_y\widehat{\psi}(y,t)$ for some scalar function $\widehat{\psi}$. Now letting $(y, f, t) = \widehat{\Psi}(y, e, t) = (y, e + \widehat{\psi}(y,t), t)$ we find that $\widehat{\Omega} = \widehat{\Psi}_*\Omega_{can}^{ext}$. Hence $\Phi = \widehat{\Psi} \circ \widehat{\Phi}$ provides the desired transformation. $\qquad\square$

Remark: From the construction process and the fact that the functions ϕ can be chosen such that $\phi(x_1,t) = x_1 + \mathcal{O}(|x_1|^3)$ we see that it is possible to have $\psi(x_1,t) = \mathcal{O}(|x_1|^4)$.

The transformation Φ retains the structure of the extended Hamiltonian system, viz. $\widetilde{H}^{ext}(x_1, e, t) = \widetilde{H}(x_1,t) - e$ tranforms into $\overline{H}^{ext}(y, f, t) = \widetilde{H}(\phi^{-1}(y,t),t) - f - \widehat{\psi}(y,t)$. Note that the contribution through $\widehat{\psi}$ does not depend on $\widetilde{H}$ but is derived from the symplectic structure only. The system in y–coordinates is now

$$\dot{y} = J_{can}\mathbf{d}_y\overline{H}(y,t), \tag{7.4}$$

where $\overline{H}(y,t) = \overline{H}^{ext}(y,0,t)$.

Since the mapping $\Phi(\cdot,\cdot,t_0)$ can be chosen to depend smoothly and uniquely on $\widetilde{\Omega}^{ext}(\cdot,t_0)$ and $\frac{d}{dt}\widetilde{\Omega}^{ext}(\cdot,t_0)$ we find that Φ has the same qualitative time–dependence as $\widetilde{\Omega}$, which in turns inherits the t–dependence from the full system (7.1) (see [Mi86a]). Thus, we have

Corollary 7.2

If the Hamiltonian $H = H(x,t)$ in (7.1) is periodic, quasi–periodic, asymptotically periodic or asymptotically constant in t, then the Hamiltonian $\overline{H} = \overline{H}(y,t)$ in (7.4) has the same property.

The addition of $-\widehat{\psi}(y,t)$ to the new Hamiltonian deviates from the general rule for autonomous systems that Hamiltonians transform just by composition. One might ask whether it is really necessary to have this correction term. Since it is desirable to keep the time–variable unchanged, we want to restrict our attention to time–preserving transformations.

For constructing these we proceed as in the proof of Theorem 7.1 and take any smooth family of mappings $y = \phi(\cdot,t)$ such that $\widetilde{\Omega}_1$ transforms into the canonical one. Thus we obtain a symplectic form $\widehat{\omega}_{(y,e,t)}$ which deviates from the canonical one only through the components $\widehat{r}(y,t) = -\mathbf{d}_y\widehat{\psi}(y,t)$. If we now use the classical way of constructing canonical coordinates (see Theorem 5.12), we have to find a one–form α sucht that $\mathbf{d}\alpha = \omega_{can} - \widehat{\omega}$. In our case $\alpha(u,v,w) = -\widehat{\psi}w + \mathbf{d}f[(u,v,w)]$ for an arbitrary function $f = f(y,e,t)$. For $\theta \in [0,1]$ we now define the vector field $V_\theta = (u_\theta, v_\theta, w_\theta)$ through $\omega_\theta(V_\theta,\cdot) = -\alpha$, where $\omega_\theta = \widehat{\omega} + \theta(\omega_{can} - \widehat{\omega})$. From the special structure of ω_θ we find

$$u_\theta = J_{1\,can}\mathbf{d}_y f(y,e,t), \quad w_\theta = \mathbf{d}_e f(y,e,t), \quad v_\theta = \widehat{\psi} - \mathbf{d}_t f + \theta\langle \mathbf{d}_y\widehat{\psi}, J_{1\,can}\mathbf{d}_y f\rangle.$$

We want to keep both, t and e, unchanged; hence we need $v_\theta = w_\theta = 0$. By taking f independent on e we obtain $w_\theta = 0$. If we want to choose $f = f(y,t)$ such that $v_\theta = 0$ as well, we realize that f has to satisfy a linear partial differential equation of first order:

$$\frac{\partial}{\partial t}f = \langle \mathbf{d}_y f, \theta J_{can}\mathbf{d}_y\widehat{\psi}\rangle + \widehat{\psi}. \tag{7.5}$$

The dependence on θ is no problem since f may depend also on θ. By prescribing any values for f at $t = t_0$ we can solve this equation by the method of characteristics and obtain, locally around $(y,t) = (0,t_0)$, a unique solution. Unfortunately, in general this solution can not be extended to a whole cylindrical domain around the time–axis (see the counterexample in Section 7.4). Only in the case that $\widehat{\psi}$ decays for $|t| \to \infty$ we can solve for f globally; a fact used later in the case of systems being asymptotically constant. Moreover the qualitative time–dependence of the problem is no longer maintained in solving (7.5). For instance, for t–periodic $H(x,t)$ we easily see that the periodicity carries through to $\widehat{\psi}$; but this does not imply that there is a t–periodic f. This is explicitly worked out in the example in Section 7.4.

Nevertheless, using the solution $f_\theta(y,t)$ the vector field $V_\theta = (u_\theta, 0, 0)$ defines locally the flow operator Ψ_θ, $\theta \in [0,1]$, $\Psi_0 =$identity; and Ψ_1^{-1} transforms $\widehat{\omega}^{ext}$ into ω_{can}. Thus we have

Proposition 7.3

Let $\omega_{(x,e,t)}^{ext}$ be given as in (7.3). If $\eta = \eta(x_1)$ transforms $\widetilde{\Omega}_1(\cdot, t_0)$ into Ω_{can} then there is a neighborhood $U_1 \subset X_1$ of zero, an $\varepsilon > 0$, and a unique function $\phi : U_1 \times (t_0 - \varepsilon, t_0 + \varepsilon) \to X_1$ with $\phi(x_1, t_0) = \eta(x_1)$, such that the mapping $\overline{\Phi} : (x_1, e, t) \to (\phi(x_1, t), e, t)$ transforms $\widetilde{\Omega}^{ext}$ into Ω_{can}^{ext}.

7.3 Piecewiese autonomous systems

A special situation arises when the Hamiltonian $H = H(x,t)$ is time–independent on some time–interval (t_0, t_1). In the example in Chapter 8 we will be especially interested in the case that H has the form $H(x,t) = H_a(x) + \varepsilon H_p(x,t)$, where the perturbation H_p vanishes outside of a bounded t–interval. Generally the reduction function $x_2 = h(x_1, t)$ will also be time–dependent on the interval (t_0, t_1). See [Mi86a] for a general theory and also the example in the next section. Hence, $\widetilde{J}$ and $\widetilde{H}$ will depend on t as well. However considering any solution $x = x(t)$ on the center manifold we have $\widetilde{H}(x_1(t), t) = H(x_1(t), h(x_1(t), t), t) =$ const. for $t \in (t_0, t_1)$. However since $\widetilde{J}_1$ in (7.2) is skew–symmetric we have

$$0 = \frac{\partial}{\partial t}H = \frac{d}{dt}\widetilde{H}(x_1(t), t) = \frac{\partial}{\partial t}\widetilde{H}(x_1, t) + \langle \mathbf{d}_{x_1}\widetilde{H}, \widetilde{J}(x_1, t)\mathbf{d}_{x_1}\widetilde{H}\rangle = \frac{\partial}{\partial t}\widetilde{H}(x_1, t).$$

Thus, we have proved

Theorem 7.4

Let $\frac{\partial}{\partial t}H(x,t) = 0$ for all $x \in X$ and all $t \in (t_0, t_1)$. Then for the reduced Hamiltonian, the relation $\frac{\partial}{\partial t}\widetilde{H}(x_1, t) = 0$ holds for all $t \in (t_0, t_1)$ and x_1 with $\|x_1\| < \varepsilon$.

To obtain again a canonical Hamiltonian system we have introduced the transformation $(x_1, t, \widetilde{H}) \to (y, t, \overline{H}) = (\phi(x_1, t), t, H(x_1, t) - \psi(x_1, t))$. Unfortunately this new Hamiltonian $\overline{H}(y,t)$ will no longer be t–independent on the interval (t_0, t_1). Only the transformation derived in Proposition 7.3 retains the property stated in Theorem 7.4.

Nevertheless, the function I defined by $I(y,t) = \widetilde{H}(\phi^{-1}(y,t), t)$ is constant along solution as long as $t \in (t_0, t_1)$. Thus we are in the strange situation of having a time–

dependent Hamiltonian system with Hamiltonian $\overline{H}(y,t)$ which has, on a certain time–interval, a first integral $I(y,t)$ such that the difference $\overline{H} - I$ is very small, viz. order $\mathcal{O}(|y|^4)$.

For the case that the time–dependence occurs only on a compact domain, let us say for $t \in (-T,T)$, it is shown in [Mi86a] that the center manifold decays exponentially for $|t| \to \infty$ against the center manifold for the corresponding autonomous system, i.e.

$$\|h(\cdot,t) - h_a(\cdot)\|_{C^2(U_1,X_2)} \leq C\,e^{-d|t|}, \ \|h_t(\cdot,t)\|_{C^1(U_1,X_2)} \leq C\,e^{-d|t|}$$

where C and d are suitable positive constants. Going through the proof of Proposition 7.3 we see that similar decay estimates hold for the function $\widehat{\psi}$ in (7.5).

But now we can show that the solution f of (7.5) also has the property that it attains a constant limit at $+\infty$ as well as $-\infty$. Therefore we note that the characteristic curves for the systems are exactly the solutions of the Hamiltonian system $\dot{y} = \theta J_{can}\mathbf{d}_y\widehat{\psi}(y,t)$. For the corresponding flow operator Π_t with $\Pi_0 =$ identity, we find

$$\|\Pi_t(\cdot) - \Pi_\infty\|_{C^1(U_1,X_1)} + \|\Pi_{-t}(\cdot) - \Pi_{-\infty}\|_{C^1(U_1,X_1)} \leq C\,e^{-dt}$$

for $t > 0$. In particular, Π_t exists on a whole cylinder $U_1 \times \mathbb{R}$. A solution f of (7.5) is now found by solving

$$\frac{\partial}{\partial t}g_\theta(\eta,t) = \widehat{\psi}(\Pi_t(\theta,\eta),t), \qquad g_\theta(\eta,0) = 0,$$

and letting $f_\theta(y,t) = g_\theta(\Pi_t^{-1}(\theta,y),t)$, where $\eta = \Pi_t^{-1}(\theta,y)$ is the inverse of $\Pi_t(\theta,\cdot)$. Thus we are able to construct a transformation $(x_1,e,t) \to (y,e,t) = (\phi(x_1,t),e,t)$ such that the new coordinates are canonical. Moreover we have not changed the values of e and t, and the new Hamiltonian is given by $\overline{H}(y,t) = \widetilde{H}(\phi^{-1}(y,t),t)$. Hence, appealing to the previous theorem we can conclude as follows:

Theorem 7.5

Let $H = H(x,t)$ be time-independent for $|t| \geq T$ for some $T > 0$. Then for the reduced system (7.2) on the center manifold there are canonical coordinates $(q,p) = \phi(x_1,t)$ such that the Hamiltonian function $\overline{H}(q,p,t)$ is time–independent for $|t| \geq T$.

Remark: It is not necessary to assume that that H has the same values for $t > T$ and $t < T$. But even if it is the case we cannot conclude that the same is true for $\overline{H}$. Therefore additional symmetry properties in t would be needed, for instance reversibility.

7.4   A four–dimensional example

To illustrate the developed theory let us study the Hamiltonian system given by

$$H(q,p,Q,P,t) = \frac{1}{2}p^2 + QP + q^2(\alpha(t)Q + \beta(t)P).$$

Here the origin is an equilibrium with a t–independent linearized flow having a double zero eigenvalue with generalized eigenspace (q, p) and the eigenvalue pair ± 1 with eigenspace (Q, P). The functions α and β are assumed to be continuously differentiable and bounded together with their derivatives.

The center manifold has the expansion

$$\begin{pmatrix} Q \\ P \end{pmatrix} = \begin{pmatrix} Q(q, p, t) \\ P(q, p, t) \end{pmatrix} = \begin{pmatrix} \gamma_0 q^2 + \gamma_1 qp + \gamma_2 p^2 \\ \delta_0 q^2 + \delta_1 qp + \delta_2 p^2 \end{pmatrix} + \mathcal{O}(3).$$

Here and in the following greek letters indicate coefficients depending on t. Inserting this into the differential equation

$$\dot{q} = p, \qquad \dot{p} = -2q(\alpha Q + \beta P),$$
$$\dot{Q} = Q + \beta q^2, \quad \dot{P} = -(P + \alpha q^2),$$

we find, by comparison of equal powers, the relations

$$\dot{\gamma} = \begin{pmatrix} 1 & 0 & 0 \\ -2 & 1 & 0 \\ 0 & 1 & 1 \end{pmatrix} \gamma + \begin{pmatrix} \beta \\ 0 \\ 0 \end{pmatrix}, \quad \dot{\delta} = \begin{pmatrix} -1 & 0 & 0 \\ -2 & -1 & 0 \\ 0 & -1 & -1 \end{pmatrix} \delta + \begin{pmatrix} -\alpha \\ 0 \\ 0 \end{pmatrix}. \tag{7.6}$$

The reduced Hamiltonian satisfies

$$\begin{aligned}
\widetilde{H}(q, p, t) &= \frac{1}{2}p^2 + (\gamma_0 \delta_0 + \alpha\gamma_0 + \beta\delta_0)q^4 + (\gamma_0 \delta_1 + \gamma_1 \delta_0 + \alpha\gamma_1 + \beta\delta_1)q^3 p \\
&\quad + (\gamma_0 \delta_2 + \gamma_1 \delta_1 + \gamma_2 \delta_0 + \alpha\gamma_2 + \beta\delta_2)q^2 p^2 \\
&\quad + (\gamma_1 \delta_2 + \gamma_2 \delta_1)qp^3 + \gamma_2 \delta_2 p^4 + \mathcal{O}(5).
\end{aligned}$$

To confirm the validity of Theorem 7.4 we assume $\dot{\alpha} = \dot{\beta} = 0$ at $t = t_0$. Then the t–derivative at $t = t_0$ of the coefficient of q^4 is $\dot{\gamma}_0 \delta_0 + \gamma_0 \dot{\delta}_0 + \alpha\dot{\gamma}_0 + \beta\dot{\delta}_0 = \dot{\gamma}_0(\delta_0 + \alpha) + \dot{\delta}_0(\gamma_0 + \beta) = \dot{\gamma}_0(-\dot{\delta}_0) + \dot{\delta}_0 \dot{\gamma}_0 = 0$. The same holds for the other coefficients.

The reduced extended symplectic structure satisfies

$$\widetilde{\Omega}^{ext}(q, p, t) = \begin{pmatrix} 0 & 1+s & 0 & r_1 \\ -1-s & 0 & 0 & r_2 \\ 0 & 0 & 0 & 1 \\ -r_1 & -r_2 & -1 & 0 \end{pmatrix} \qquad \text{with}$$

$$\begin{aligned}
s &= D_q Q D_p P - D_p Q D_q P = \sigma_0 q^2 + \sigma_1 qp + \sigma_2 p^2 + \mathcal{O}(3), \\
r_1 &= D_q Q \dot{P} - D_q P \dot{Q} = \rho_0 q^3 + \rho_1 q^2 p + \rho_2 qp^2 + \rho_3 p^3 + \mathcal{O}(4), \\
r_2 &= D_p Q \dot{P} - D_p P \dot{Q} = \tfrac{1}{3}(\rho_1 - \dot{\sigma}_0)q^3 + \tfrac{1}{2}(2\rho_2 - \dot{\sigma}_1)q^2 p + (3\rho_3 - \sigma_2)qp^2 + \rho_4 p^3 + \mathcal{O}(4).
\end{aligned}$$

The special form of the coefficients in r_2 is a consequence of the closedness condition $d\widetilde{\omega}^{ext} = 0$ which reduces to $\dot{s} = D_p r_1 - D_q r_2$. Of course, σ_j and ρ_j can be expressed explicitly through γ and δ.

Now employing the transformation

$$\phi(q,p,t) = \begin{pmatrix} q \\ p \end{pmatrix} + \begin{pmatrix} \mu_0 q^3 + \ldots + \mu_3 p^3 \\ \nu_0 q^3 + \ldots + \nu_3 p^3 \end{pmatrix} + \mathcal{O}(4)$$

we find $D\phi^* \widetilde{\Omega}_1 D\phi = \widehat{\Omega}_1 = \begin{pmatrix} 0 & 1 \\ -1 & 0 \end{pmatrix}(1 + \widehat{s})$ and $\widehat{r} = r + D\phi^* \widetilde{\Omega}_1 \dot{\phi}$ with

$$\widehat{s} = (\sigma_0 + 3\mu_0 + \nu_1)q^2 + (\sigma_1 + 2\mu_1 + 2\nu_1)qp + (\sigma_2 + \mu_2 + 3\nu_3)p^2 + \mathcal{O}(3),$$
$$\widehat{r}_1 = (\dot{\nu}_0 + \rho_0)q^3 + \ldots + (\dot{\nu}_3 + \rho_3)p^3 + \mathcal{O}(4),$$
$$\widehat{r}_2 = (\tfrac{1}{3}(\rho_1 - \dot{\sigma}_0) - \dot{\mu}_0)q^3 + \ldots + (\rho_4 - \dot{\mu}_3)p^3 + \mathcal{O}(4).$$

To have $\widehat{\Omega}^{ext}$ canonical up to terms of order $\mathcal{O}(3)$ we need $\widehat{s} = \mathcal{O}(3)$ and $\widehat{r} = \mathcal{O}(4)$. The first relation is achieved by expressing μ_0, μ_1, and μ_2 in terms of ν_1, ν_2, and ν_3 and the coefficients in s. Then we are left with the eqution $\widehat{r} = \mathcal{O}(4)$. Equating the coefficients equal to zero is equivalent to

$$\dot{\mu}_3 = \rho_4 \quad \text{and} \quad \dot{\nu}_j = -\rho_j, \quad \text{for } j = 0, \ldots, 3.$$

Hence, prescribing initial conditions at $t = t_0$ we find a unique solution $(\mu, \nu)(t)$ such that $\widehat{\Omega}^{ext}$ is canonical up to third order. This corresponds to the result of Proposition 7.3.

Yet, if the coefficients μ and ν are not bounded over $t \in \mathbb{R}$, the transformation $\phi = \phi(q, p, t)$ cannot be defined uniformly over a cylindrical domain around the t–axis. We show that unbounded coefficients can occur by considering an example. Letting $\alpha = \beta = 2\cos t$ we find $\gamma_0 = \sin t - \cos t$ and $\delta_0 = -(\sin t + \cos t)$. From $\rho_0 = 2(\gamma_0 \dot{\delta}_0 - \dot{\gamma}_0 \delta)$ we obtain $\rho_0 \equiv 4$, and hence $\nu_0(t) = -4t + \nu_0(0)$. On the one hand this provides an example with ν unbounded, and on the other hand we also see that ν is not periodic although the problem is periodic. According to Corollary 7.2 we can find a periodic transformation $\phi(q, p, t)$ but only at the expense of non–vanishing $\widehat{r}$. For the third order terms we can simply take $\mu = -(\tfrac{1}{3}\sigma_0, \tfrac{1}{2}\sigma_1, \sigma_2, 0)$ and $\nu = 0$.

Part II

Applications to variational problems

Chapter 8

Elliptic variational problems on cylindrical domains

In many physical situations, such as deformations of beams, inviscid channel flows, or diffusion in pipes, one has to study variational problems in cylindrical domains. This leads to elliptic systems which recently have been studied by so–called dynamical methods. This means that the axial variable takes formally the role of a time–variable. Although the initial value problem for the corresponding differential equation is not well–posed, certain methods, originally developed for dynamical systems only, have been generalized to fit for this situation. In particular we refer to the center manifold theory, first introduced for elliptic problems by Kirchgässner [Ki82] and further developed in [Mi90, IV91]. Some of the problems discussed there will be taken up once again to show how the variational structure, previously not exploited, enters into the analysis.

In applications there exist well–established reductions methods for Lagrangian problems, which are obtained from elliptic variational problems by treating the axial variable as time–like. But these projection or Galerkin methods are only approximative with more or less accuracy depending on the physically motivated ansatz. For an example in fluid dynamics see [BB87]. In [An72] and [Ow87] the projection method was used in elasticity theory in order to reduce the partial differential equations of three–dimensional beam theory to the ordinary differential equations of rod theory. It is the aim of this work to lay a rigorous basis for a reduction of this kind.

The Hamiltonian structure, we associate to elliptic variational problems, can also be used independently of the reduction context. For instance, often one is interested in the bifurcations of (quasi–) periodic, homoclinic or heteroclinic solutions with respect to the axial variable of the cylinder. The Hamiltonian structure can now be exploited to explain why certain bifurcations happen while others can not appear. In [BM90] this approach is used for a problem in the theory of water waves.

8.1   Second order elliptic systems

We consider the cylindrical domain $\Omega = I \times \Sigma$ where $I \subset \mathbb{R}$ is a possibly infinite interval and the cross–section $\Sigma \subset \mathbb{R}^n$ is a bounded domain with smooth boundary. We denote the axial and the cross–sectional variables by t and y, respectively; and let $x = (t, y)$ with $(x_0, x_1, \ldots, x_n) = (t, y_1, \ldots, y_n)$ whenever convenient. Note that t still is a spatial variable but it will play formally the role of a time–variable. For a function $u : \Omega \to \mathbb{R}^m$ we write ∇u for the $n \times m$–matrix consisting of the partial derivatives with respect to $(y_1, \ldots, y_m)$ and let $\nabla_x u = (\dot{u}, \nabla u)$.

Given functions $f = f(y, u, \nabla_x u)$ and $g = g(y, u)$ we define the functional

$$\mathcal{I}(u) = \int_\Omega f(y, u, \nabla_x u) \, dy + \int_{\partial\Omega} g(y, u) \, dy. \tag{8.1}$$

The aim is to find a u in a certain subset V of $W^{1,p}(\Omega)$ such that $\mathcal{I}$ attains a local extremum in u or, at least, makes $\mathcal{I}$ stationary (i.e. $\delta\mathcal{I}(u) = 0$). We assume that the subset V is characterized by boundary conditions only:

$$V = \{\, u \in W^{1,p}(\Omega) \;:\; u(t, y) = \phi(y) \text{ on } I \times \Gamma_0 \subset \partial\Omega \,\}.$$

Here Γ_0 is a smooth part of the boundary $\partial\Sigma$. For simplicity we do not include explicit dependence on the axial variable t. Using the methods of Chapter 7 it is obvious how this case can be treated analogously. Without loss of generality we let $\phi = 0$ since u can be changed into $u + \tilde{u}$ with $\tilde{u}|_{I \times \Gamma_0} = \phi$. We will not care for boundary conditions on the terminal surfaces $\partial I \times \Sigma$ since they do not enter our analysis. The center manifold theory is actually concerned with solutions on the infinite cylinder $\Omega = \mathbb{R} \times \Sigma$. However, using the abstract Saint–Venant's principle as introduced in [Mi90] it follows that every small solution on a long but finite cylinder can be very well approximated by a solution on the center manifold. A general solution can be modelled as a solution on the center manifold plus boundary layers which decay exponentially with the distance from the terminal ends.

The first variation of $\mathcal{I}$ has the form

$$\delta\mathcal{I}(u)[v] = \int_\Omega \left(\sum_{j=1}^m \frac{\partial}{\partial u_j} f(y, u, \nabla_x u) v_j + \sum_{j=1}^m \sum_{i=0}^n \frac{\partial}{\partial u_{j,i}} f(y, u, \nabla_x u) v_{j,i} \right) dy$$

$$+ \int_{I \times \Gamma_1} \sum_{j=1}^m \frac{\partial}{\partial u_j} g(y, u) v_j \, dy$$

Here we use the notation $u_{j,i} = \frac{\partial}{\partial x_i} u_j$ and $\Gamma_1 = \partial\Sigma \backslash \Gamma_0$. The boundary integral over $I \times \Gamma_0$ can be omitted as v has to vanish there. For smooth u we have $\delta\mathcal{I}(u) = 0$ if and only if u satisfies the the *Euler–Lagrange equation*, which is obtained by partial integration:

$$-\sum_{i=0}^n \frac{\partial}{\partial x_i} \left(\frac{\partial}{\partial u_{j,i}} f(y, u, \nabla_x u) \right) + \frac{\partial}{\partial u_j} f(y, u, \nabla_x u) = 0 \tag{8.2}$$

for $j = 1, \ldots, m$. Additionally, we have the boundary conditions

$$u = 0 \quad \text{on } I \times \Gamma_0, \qquad \sum_{i=1}^{n} \frac{\partial}{\partial u_{j,i}} f(y, u, \nabla_x u) \tilde{n}_i = \frac{\partial}{\partial u_j} g(y, u) \quad \text{on } I \times \Gamma_1, \qquad (8.3)$$

where $\tilde{n} = (\tilde{n}_1, \ldots, \tilde{n}_n)$ is the unit outward normal vector at $\partial \Sigma$.

For variational problems it is classical to assume that the system (8.2) is elliptic in each point $(y, u, \nabla_x u)$ of interest. To define ellipticity we write (8.2) in the form

$$\sum_{j=1}^{m} \sum_{k,l=0}^{n} c_{ijkl} u_{j,kl} + \text{ lower order terms } = 0, \qquad (8.4)$$

where $c_{ijkl} = \frac{\partial^2}{\partial u_{i,k} \partial u_{j,l}} f(y, u, \nabla_x u)$. Note that $c_{ijkl} = c_{jikl} = c_{ijlk}$ for all i, j, k, and l. The Euler–Lagrange equation is called *elliptic* in a point $(y, u, \nabla_x u)$ if

$$\sum_{i,j=1}^{m} \sum_{k,l=0}^{n} c_{ijkl}(y, u, \nabla_x u) \eta_i \eta_j \xi_k \xi_l \neq 0 \qquad (8.5)$$

for all $\eta = (\eta_1, \ldots, \eta_m) \neq 0$ and all $\xi = (\xi_0, \ldots, \xi_n) \neq 0$. Furthermore, (8.2) is called *strongly elliptic*, if there is a positive ε such that

$$\sum_{i,j=1}^{m} \sum_{k,l=0}^{n} c_{ijkl}(y, u, \nabla_x u) \eta_i \eta_j \xi_k \xi_l \geq \varepsilon |\eta|^2 |\xi|^2 \qquad (8.6)$$

for all η and ξ. This condition is also called *rank–one convexity*, since it implies that

$$f(y, u, tF + (1-t)G) \leq t f(y, u, F) + (1-t) f(y, u, G) \qquad \text{for all } t \in [0,1],$$

whenever $F - G$ is a $(n+1) \times m$–matrix of rank one, viz. $F - G = \eta \otimes \xi$.

In [Ba77, Da89] more restrictive convexity properties on f are given which are sufficient to guarantee the existence of minimizers when combined with appropriate coercivity conditions. We do not need any of these results since our interest is restricted to local considerations close to a t–independent solution. But it turns out that ellipticity is strong enough to enable us to tansfer the variational problem into a Hamiltonian one.

First we formulate the variational problem as a Lagrangian problem by introducing the spaces $Q = Q = L_2(\Sigma, \mathbb{R}^m)$, $Q_1 = H^k(\Sigma, \mathbb{R}^m)$, and $Q_2 = \{ u \in H^{k+1}(\Sigma, \mathbb{R}^m) : u|_{\Gamma_0} = 0 \}$. Here, $H^k(\Sigma, \mathbb{R}^m)$ is the classical Sobolev space containing those functions having all derivatives up to order k in $L_2(\Sigma, \mathbb{R}^m)$. Furtheron we drop the argument $\mathbb{R}^m$. Obviously Q_1 and Q_2 are dense in $L_2(\Sigma)$ with continuous embeddings. Hence, $\mathcal{Y} = Q_2 \times Q_1$ is a manifold domain in $TQ = Q \times Q$.

We take $k > n/2$, then Sobolev's embedding theorem tell us that $H^k(\Sigma)$ is continuously embedded in $C^0(\overline{\Sigma})$. Now the Lagrangian

$$L(u, \dot{u}) = \int_\Omega f(y, u, \nabla u, \dot{u})\, dy + \int_{\Gamma_1} g(y, u)\, dy$$

is a smooth function on $\mathcal{Y}$. Obviously, the Lagrange equation corresponding to L is exactly the Euler–Lagrange equation corresponding to $\mathcal{I} = \int_I L\, dt$.

The fiber derivative FL maps $(u, \dot{u})$ into $(u, v) = (u, \frac{\partial}{\partial \dot{u}} f(y, u, \nabla u, \dot{u}))$. However, the ellipticity condition, where ξ is to be taken as $(1, 0, \dots, 0)$, implies that the $m \times m$–matrix $\frac{\partial^2}{\partial \dot{u}^2} f(\cdots)$ is invertible. Thus, the implicit function theorem yields locally an inverse function

$$\dot{u} = r(y, u, \nabla u, v).$$

If even the strong ellipticity condition (8.6) holds for all $(y, u, \nabla_x u)$, then $\frac{\partial^2}{\partial \dot{u}^2} f(\cdots)$ is positive definite; and hence $f(y, u, \nabla u, \cdot)$ is convex in $\dot{u}$ for each fixed $(y, u, \nabla u)$. The classical Legendre transform gives now the function r globally:

$$r(y, u, \nabla u, v) = \max_{\dot{u} \in \mathbf{R}^m} [\dot{u} \cdot v - f(y, u, \nabla u, \dot{u})].$$

In any case these pointwise functions show that the fiber derivative is locally regular (globally if (8.2) is strongly elliptic). Therefore note that $\frac{\partial f}{\partial \dot{u}}$ as well as r are locally smooth functions from $Q_2 \times Q_1 = Q_2 \times Q_1^*$ into itself.

Using the fiber derivative FL we can define the action A, the energy E, and the Hamiltonian H. The last one takes the form

$$H(u, v) = \int_\Omega h(y, u, \nabla u, v)\, dy - \int_{\Gamma_1} g(y, u)\, dy \qquad (8.7)$$

where $h(y, u, \nabla u, v) = v \cdot r(y, u, \nabla u, v) - f(y, u, \nabla u, r(\cdots))$. The basic manifold remains $T^*Q = L_2(\Sigma) \times L_2(\Sigma)$ with its canonical symplectic form. H is a smooth function on the manifold domain $\mathcal{Y}^* = Q_2 \times Q_1^*$.

Applying the results provided above in the abstract setting, we are now able to conclude that the reduction of a variational elliptic problem onto its center manifold M_C leads to a reduced Hamiltonian system. By studying the linearized problem Theorem 6.9 allows us to decide whether the full nonlinear system is locally generated by a Lagrangian. In practice, the generalized eigenvectors for the eigenvalues on the imaginary axis have to be computed for the approximation of the center manifold; hence it is no additional effort to check the assumptions of Theorem 6.5. Then we find coordinates $(\tilde{q}, \dot{\tilde{q}})$ such that M_C is given by $u = r(\tilde{q}, \dot{\tilde{q}})$, $\quad \dot{u} = s(\tilde{q}, \dot{\tilde{q}})$. Moreover, we obtain a reduced Lagrangian $\tilde{L}(\tilde{q}, \dot{\tilde{q}})$ such that the variational problem corresponding to the functional

$$\tilde{\mathcal{I}}(\tilde{q}) = \int_I \tilde{L}(\tilde{q}, \dot{\tilde{q}})\, dt$$

is equivalent to the variational problem for $\mathcal{I} = \mathcal{I}(u)$ defined in (8.1), in the sense of center manifold reduction.

8.2   A first example

The simplest example is given by $f(y, u, \nabla_x u) = \frac{1}{2}(\dot{u}^2 + (\partial u/\partial y)^2 + \mu u^2) + b(u)$ with $b(u) = \beta u^3 + \mathcal{O}(u^4)$, $\Sigma = (0, \pi) \subset \mathbb{R}$, and $\Gamma_0 = \{0, \pi\}$. Because of the simple form we find $\dot{u} = r(y, u, \nabla u, v) = v$ implying that the corresponding Lagrangian L and the associated Hamiltonian H take the form

$$L(u, \dot{u}) = \int_0^\pi \left[\frac{1}{2}(\dot{u}^2 + (\frac{\partial}{\partial y}u)^2 + \mu u^2) + b(u) \right] dy,$$

$$H(u, v) = \int_0^\pi \left[\frac{1}{2}(\dot{u}^2 - (\frac{\partial}{\partial y}u)^2 - \mu u^2) - b(u) \right] dy.$$

Both, the Lagrange and the Hamilton equation can be written as

$$\frac{d}{dt}\begin{pmatrix} u \\ v \end{pmatrix} = \begin{pmatrix} 0 & I \\ -\frac{\partial^2}{\partial y^2} + \mu & 0 \end{pmatrix} \begin{pmatrix} u \\ v \end{pmatrix} + \begin{pmatrix} 0 \\ b'(u) \end{pmatrix} = K\begin{pmatrix} u \\ v \end{pmatrix} + N(u) \tag{8.8}$$

where $\binom{u}{v}$ can be understood as elements of $Z = H_0^1(0, \pi) \times L_2(0, \pi)$ and where $D(K) = (H^2(0, \pi) \cap H_0^1(0, \pi)) \times H_0^1(0, \pi)$. Equations of this type are studied in [Ki82, Mi86a] where also proofs for the existence of the center manifold are given. We motivate the center manifold approach a little further in mentioning that system (8.8) can also be written as an infinite system of coupled ordinary differential equations as follows. We insert $u(y, t) = \sum_1^\infty r_n(t) \sin ny$ into the equations and obtain, by projection onto $\sin ny$,

$$\ddot{r}_n - (n^2 + \mu)r_n = f_n((r_k)_{k \in N}), \quad n = 1, 2, \ldots$$

For $n^2 + \mu > 0$ the left hand side of this equation can be inverted on the space of slowly exponentially growing function:

$$r_n(t) = \int_{s \in R} -\frac{1}{2\tau_n} e^{-\tau_n |t-s|} f_n((r_k(s)) \, ds, \quad \tau_n = \sqrt{n^2 + \mu}.$$

Putting all these components together, we obtain the Green's function G_2 introduced in (2.9). Doing the functional analysis in the above given spaces we realize that the center manifold theorem is applicable.

Note that the operator K has the eigenvalues $\pm(n^2 + \mu)^{1/2}$ with the corresponding eigenfunctions $\phi_{\pm n} = (\sin ny, \pm(n^2 + \mu)^{1/2} \sin ny)^T$ if $n^2 + \mu \neq 0$. Hence for $\mu \leq -1$ there is a nontrivial center space. To avoid unnecessarily technical notations we just treat one typical case, let us say $\mu = -9$. All the other cases follow similarly. Now the center manifold is six–dimensional; we have two complex pairs, $\pm i\sqrt{8}$ and $\pm i\sqrt{5}$, with one–dimensional eigenspaces. Additionally we have a two–fold zero eigenvalue with a generalized kernel spanned by $e_1 = (\sin 3y, 0)^T$ and $e_2 = (0, \sin 3y)^T$. From $Ke_2 = e_1$ we see that this corresponds to **Case 3** with $m = 2$ in the normal form theory. Hence, Theorem 6.5 guarantees that the flow on the center manifold is a Lagrangian flow.

To calculate the reduced problem we choose a symplectic basis in the tangent space of the center manifold in $(u, v) = 0$. The canonical symplectic form on $L_2(0, \pi) \times L_2(0, \pi)$ is given by $\omega_{can}((u_1, v_1), (u_2, v_2)) = \int_0^\pi (u_1 v_2 - v_1 u_2)\, dy$. Hence, the vectors $\sqrt{2/\pi}(\sin ny, 0)^T$ and $\sqrt{2/\pi}(0, \sin ny)^T$, $n = 1, 2, 3$, form the desired basis, and the center manifold can be given by

$$\begin{pmatrix} u \\ v \end{pmatrix} = \sum_{n=1}^{3} \sqrt{\frac{2}{\pi}} \left(r_n \begin{pmatrix} \sin ny \\ 0 \end{pmatrix} + s_n \begin{pmatrix} 0 \\ \sin ny \end{pmatrix} \right) + h(r, s).$$

Using Theorem 4.2 and $b'(u) = 3\beta u^2 + \mathcal{O}(|u|^3)$ the reduced Hamiltonian takes the form

$$\begin{aligned} \widetilde{H}(r, s) ={}& \tfrac{1}{2}(s_1^2 + s_2^2 + s_3^2 + 8r_1^2 + 5r_2^2 + 0r_3^2) \\ &- \left(\tfrac{2}{\pi}\right)^{3/2} \beta \left(\tfrac{4}{3}r_1^3 + \tfrac{4}{9}r_3^3 + \tfrac{16}{5}r_1 r_2^2 + \tfrac{108}{35}r_1 r_3^2 - \tfrac{4}{5}r_1^2 r_3 + \tfrac{16}{7}r_2^2 r_3 \right) + \mathcal{O}(|(r, s)|^4). \end{aligned}$$

Since the reduced symplectic structure $\widetilde{\Omega}$ satisfies $\widetilde{\Omega}(r, s) = \left(\begin{smallmatrix} 0 & I \\ -I & 0 \end{smallmatrix}\right) + \mathcal{O}(|(r, s)|^2)$ we find, according to Theorem 5.15, canonical coordinates $(q, p) = (r, s) + \mathcal{O}(|(r, s)|^3)$. Hence, giving $\widetilde{H}$ in terms of (q, p) yields the same expansion. In particular, the Legendre transformation gives $\dot{q} = p + \mathcal{O}(|(q, p)|^3)$; and for the reduced Lagrangian we obtain

$$\begin{aligned} \widetilde{L}(q, \dot{q}) ={}& \tfrac{1}{2}(\dot{q}_1^2 + \dot{q}_2^2 + \dot{q}_3^2 - 8q_1^2 - 5q_2^2 - 0q_3^2) \\ &+ \left(\tfrac{2}{\pi}\right)^{3/2} \beta \left(\tfrac{4}{3}q_1^3 + \tfrac{4}{9}q_3^3 + \tfrac{16}{5}q_1 q_2^2 + \tfrac{108}{35}q_1 q_3^2 - \tfrac{4}{5}q_1^2 q_3 + \tfrac{16}{7}q_2^2 q_3 \right) + \mathcal{O}(|(q, \dot{q})|^4). \end{aligned}$$

The center manifold has the form

$$u = \sum_{n=1}^{3} q_n \sqrt{\frac{2}{\pi}} \sin ny + m_1(q, \dot{q}), \qquad \dot{u} = \sum_{n=1}^{3} \dot{q}_n \sqrt{\frac{2}{\pi}} \sin ny + m_2(q, \dot{q}), \qquad (8.9)$$

where $(m_1, m_2) = \mathcal{O}(|(q, \dot{q})|^2)$ and $\int_0^\pi m_i \sin ny\, dy = 0$ for $i = 1, 2$ and $n = 1, 2, 3$. Inserting this into the original Lagrangian L we find that the projected Lagrangian

$$L_P(q, \dot{q}) = L(u(q, \dot{q}), \dot{u}(q, \dot{q}))$$

has exactly the same expansion as $\widetilde{L}$, up to terms of fourth order. Hence, in this case the reduction and the projection method coincide at least for the lowest order terms. This means that, for the lowest order terms, we have achieved *natural reduction* in the sense of Section 6.5.

8.3   A nonautonomous example

We consider internal waves of a stratified fluid in a two–dimensional horizontal channel given by $\Omega = \{ (t, y) \in I\!R^2 : t \in I\!R, \; y \in (g_1(t), g_2(t)) \}$, see [Ki82, Mi86b] for the physical motivation. Let ρ be the density and (u, v) the velocity, then the pseudo stream function ψ defined by $\sqrt{\rho}(u, v) = (\psi_y, -\psi_t)$ satisfies the Long–Yih equation:

$$\Delta\psi + r(\lambda, \psi) = 0 \quad \text{in } \Omega, \quad \left\{ \begin{array}{c} \psi(t, g_1(t)) = 0 \\ \psi(t, g_2(t)) = \psi_0 \end{array} \right\} \text{ for } t \in I\!R, \tag{8.10}$$

where r is a known function to be calculated from the inflow conditions at $t = -\infty$. The parameter λ is dimensionless and relates the inflow velocity to the density; $F = 1/\sqrt{\lambda}$ is the Froude number.

Letting $R(\lambda, \psi) = \int_0^\psi r(\lambda, s)\, ds$ we see that (8.10) is the Euler–Lagrange equation of the functional $\mathcal{I}(\psi) = \int_\Omega \left\{ \frac{1}{2}(\psi_t^2 + \psi_y^2) - R(\lambda, \psi) \right\} dy\, dt$. To apply our methods we transform Ω into the straight strip $\Omega_0 = I\!R \times \Sigma$, $\Sigma = (0, 1)$, by setting $y = g_1(t) + \eta(g_2(t) - g_1(t))$ and $\Psi(t, \eta) = \psi(t, (y - g_1(t))/(g_2(t) - g_1(t)))$. As in [Mi86b] we assume $g_1(t) = \varepsilon g(t)$ and $g_2(t) \equiv 1$. There a conformal mapping $K : \Omega \to \Omega_0$ was used to obtain again a nonlinear Laplace equation with a (t, y)–dependent factor in front of r. The transformation used here destroys this structure, but still the existence of the center manifold can be shown by using Theorem 2.1.

Now $\mathcal{I}$ transforms into $\mathcal{I}_0(\Psi) = \int_{I\!R} L(\Psi, \Psi_t, t, \varepsilon)\, dt$ with

$$L(\Psi, \Psi_t, t, \varepsilon) = \int_{\eta=0}^1 \left\{ \frac{1}{2}\left(\Psi_t - \frac{\varepsilon(1 - \eta)g'}{1 - \varepsilon g}\Psi_\eta \right)^2 + \frac{1}{2}\left(\frac{\Psi_\eta}{1 - \varepsilon g} \right)^2 - R(\lambda, \Psi) \right\} (1 - \varepsilon g)\, d\eta.$$

According to the first section we find the associated Hamiltonian H by setting $\Phi = \partial L/\partial\Psi_t = \{\Psi_t - \varepsilon(1 - \eta)g'\Psi_\eta/(1 - \varepsilon g)\}(1 - \varepsilon g)$. It takes the form

$$H(\Psi, \Phi, t, \varepsilon) = \int_{\eta=0}^1 \left\{ \frac{1}{2(1-\varepsilon g)}\left(\Phi^2 + 2\varepsilon(1-\eta)g'\Phi\Psi_\eta - \Psi_\eta^2 \right) + (1 - \varepsilon g)R(\lambda, \Psi) \right\} d\eta.$$

We assume that the channel is flat outside the region $|t| \geq \ell$ with $g(t) = 0$, i.e. we are looking for steady flow patterns over a single bump at the bottom of the channel. Then the Hamiltonian H does no longer depend on t on these regions. Hence, H is conserved along solutions of (8.10) and Theorem 7.5 is applicable.

For $\varepsilon = 0$ there is always the trivial parallel flow given by $\Psi_0 = \Psi(\lambda, \eta)$ as solution of $\Psi_{\eta\eta} + r(\lambda, \Psi) = 0$, $\Psi(0) = 0$, and $\Psi(1) = \psi_0$. To study flows being close to the parallel flow we let $\Psi(t, \eta) = \Psi_0(\lambda, \eta) + U(t, \eta)$. The linearized problem for $\varepsilon = 0$ leads to the eigenvalue problem

$$D_\lambda U = -\frac{d^2}{d\eta^2}U - \frac{\partial}{\partial\Psi}r(\lambda, \Psi_0(\lambda, \eta))U = \mu U, \text{ for } \eta \in (0, 1); \; U(0) = U(1) = 0.$$

Any eigenvalue μ of D_λ generates the eigenvalue pair $\pm\sqrt{\mu}$ for the associated linear operator $K = JA$. Hence, we obtain a nontrivial center manifold if D_λ has negative eigenvalues. Denote the smallest eigenvalue of D_λ by $\nu(\lambda)$. In [Ki82, Mi86b], under fairly general conditions on the inflow profile, the existence of a critical value $\lambda_0 > 0$ is shown, such that $\nu(\lambda_0) = 0$ and $\frac{d}{d\lambda}\nu(\lambda_0) < 0$ holds.

We restrict our attention to the case $\lambda \approx \lambda_0$. Then the center manifold is two–dimensional and the flow on it is described by an ordinary differential equation of the form

$$\ddot{\alpha} + \nu(\lambda)\alpha + c(\lambda)\alpha^2 + \varepsilon G(t) + M(\varepsilon, \lambda, t, \alpha, \dot{\alpha}) = 0,$$

where $m(\lambda, \alpha, \dot{\alpha}) = M(0, \lambda, 0, \alpha, \dot{\alpha}) = \mathcal{O}(|\alpha|^3 + |\dot{\alpha}|^2)$, $M - m = \mathcal{O}(|\varepsilon|\,|(\varepsilon, \alpha, \dot{\alpha})|)$, and $|M - m| \leq Ce^{-d|t|}$. See [Mi86b] for the details.

However, the abstract theory of Chapter 7 tells us that this ODE actually corresponds to a canonical Hamiltonian system with a reduced Hamiltonian $\widetilde{H}$ of the form

$$\widetilde{H}(q, p, t, \lambda, \varepsilon) = \frac{1}{2}(p^2 + \nu(\lambda)q^2) + \frac{c(\lambda)}{3}q^3 + \varepsilon q G(t) + \text{h.o.t.}$$

Here the coordinates can be chosen such that $\widetilde{H}$ is independent of t for $t \geq \ell$ and for $t \leq -\ell$, respectively, due to Theorem 7.5.

This implies that the qualitative behavior of (q, p) for $t \to \infty$ and $t \to -\infty$ can be characterized completely by a phase plane analysis. In particular, solutions staying bounded have to be either constant, periodic or asymptotically constant for $t \geq \ell$. The same is true for $t \leq -\ell$. However, one and the same solution may have different behavior for $t < -\ell$ and for $t > \ell$ with different energy, which can changes inside the transient region $|t| \leq \ell$.

In particular, we are able to exclude the existence of solutions having an ω–limit set consisting of the union of the origin and the homoclinic orbit existing for $\nu(\lambda) < 0$. This answers the question left open in [Mi86b, Sect.5]. Of course, the whole discussion there can be considerably simplified by using the Hamiltonian approach developed here.

Going back to the original system, the solutions have the form $\Psi(t, \eta) = \Psi_0(\lambda, \eta) + U(q(t), p(t), t, \eta, \lambda, \varepsilon)$. Note that they will only be asymptotically periodic for exactly periodic (q, p), since U involves the reduction function which contains exponentially decaying terms.

Chapter 9

Capillarity surface waves

We consider steady waves in a two–dimensional horizontal fluid layer of height h travelling with velocity c. The fluid is inviscid and has constant density ρ with gravity g acting in vertical direction. The upper boundary is a free surface $y = z(x)$ where surface tension $\kappa > 0$ is present inducing a pressure jump proportional to the curvature of the surface.

Introducing non–dimensional variables in a coordinate system moving with the wave the problem takes the form

$$
\left.
\begin{aligned}
\Delta\phi &= 0 && \text{for} \quad x \in \mathbb{R},\ y \in (0, z(x)), \\
\frac{\partial}{\partial y}\phi &= 0 && \text{for} \quad x \in \mathbb{R},\ y = 0, \\
\frac{\partial}{\partial y}\phi - \dot{z}\frac{\partial}{\partial x}\phi &= 0 \\
\frac{1}{2}|\nabla\phi|^2 + \lambda z - b\frac{\ddot{z}}{(1+\dot{z}^2)^{3/2}} &= \frac{1}{2} + \lambda
\end{aligned}
\right\} \quad \text{for} \quad x \in \mathbb{R},\ y = z(x);
$$

$$(9.1)$$

where $\dot{\ } = \partial/\partial x$. Here ϕ is the potential function for the irrotational velocity field. The second and third equations state that the flat bottom and the surface $y = z(x)$ are streamlines. The last condition is Bernoulli's equation. The parameters λ and b are given by $\lambda = gh/c^2$ and $b = \kappa/(\rho h c^2)$. $F = \lambda^{-1/2}$ is called the Froude number, and b is called the Bond number.

We are looking for solutions being close to the parallel flow $(\phi(x,y), z(x)) = (x, 1)$. This problem was extensively studied within the last decade using center manifold theory for elliptic problems, see [AK89, IK91] and the references therein. However, only in [BM90], where the problem without surface tension $(b = 0)$ is treated, the Hamiltonian structure is introduced to explain the bifurcations of periodic solutions from the so–called Stokes family.

9.1   The variational formulation

It is shown in [BO82] that the steady Euler equations for certain inviscid flow problems are the Euler–Lagrange equations of the associated potential energy. In our case this reads $\mathcal{I}(\phi, z) = \int_{-\infty}^{\infty} \overline{L}(\phi, \dot{\phi}, z, \dot{z}) \, dx$, where the Lagrangian is given by

$$\overline{L}(\phi, \dot{\phi}, z, \dot{z}) = \int_0^z \frac{1}{2} \left(\dot{\phi}^2 + (\frac{\partial \phi}{\partial y})^2 \right) dy + \frac{\lambda}{2}(z-1)^2 - \frac{z}{2} + b(\sqrt{1 + \dot{z}^2} - 1).$$

Because of the free surface the basic manifold is $\mathcal{Q} = \{ (\phi, z) \, : \, z > 0, \phi \in L_2(0, z) \}$. The Lagrangian system is defined on a manifold domain in $T\mathcal{Q}$.

   To do explicit calculations we introduce local coordinates in $T\mathcal{Q}$ in a neighborhood of $(\phi, z) = (C, 1)$ where C is any constant. We let $y = z\eta$ and $\Phi(\eta) = \phi(z\eta)$. The derivatives of ϕ transform into

$$\dot{\phi}(x, z\eta) = \dot{\Phi}(x, \eta) - \frac{\eta \dot{z}}{z}\Phi'(x, \eta), \qquad \frac{\partial}{\partial y}\phi(x, z\eta) = \frac{1}{z}\Phi'(x, \eta),$$

where $' = \partial/\partial\eta$, and hence the new Lagrangian is

$$L(\Phi, z, \dot{\Phi}, \dot{z}) = \frac{1}{2z} \int_0^1 \left[(z\dot{\Phi} - \eta\dot{z}\Phi')^2 + \Phi'^2 \right] d\eta + \frac{\lambda}{2}(z-1)^2 - \frac{z}{2} + b(\sqrt{1 + \dot{z}^2} - 1).$$

This defines a Lagrangian system in a neighborhood of $(\Phi, z) = (C, 1)$ on the tangent bundle of the linear space $Q = L_2(0,1) \times I\!R$. Writing down the associated Euler–Lagrange equations explicitly, it can be checked that they are equivalent to the system (9.1).

   The Legendre transform yields the conjugate variables

$$\Psi = \frac{\partial L}{\partial \dot{\Phi}} = z\dot{\Phi} - \dot{z}\eta\Phi', \quad w = \frac{\partial L}{\partial \dot{z}} = \frac{b\dot{z}}{\sqrt{1 + \dot{z}^2}} - \frac{1}{z}\int_0^1 (z\dot{\Phi} - \dot{z}\eta\Phi')\eta\Phi' d\eta.$$

These relations are locally invertible giving

$$\dot{z} = \frac{\widetilde{w}}{\sqrt{b^2 - \widetilde{w}^2}}, \quad \dot{\Phi} = \frac{1}{z}\left(\Psi + \frac{\eta\widetilde{w}\Psi'}{\sqrt{b^2 - \widetilde{w}^2}} \right)$$

where $\widetilde{w} = w + \frac{1}{z}\int_0^1 \eta\Psi\Phi' d\eta$. The Hamiltonian is

$$H(\Phi, z, \Psi, w) = \frac{1}{2z}\int_0^1 (\Psi^2 - \Phi'^2)d\eta + \frac{z}{2} - \frac{\lambda}{2}(z-1)^2 + b - \sqrt{b^2 - \widetilde{w}^2}.$$

Note that the Hamiltonian is invariant under $\Phi \to \Phi + \gamma$ for any $\gamma \in I\!R$. Hence, $\int_0^1 \Phi d\eta$ is a cyclic variable and the conjugate variable $\int_0^1 \Psi d\eta$ is a first integral. In fact, $\int_0^1 \Psi d\eta = \int_0^z \dot{\phi} dy$ is exactly the mass flow through the cross-section $\{x\} \times (0, z(x))$, which has to be independent of x due to the conservation of mass. In [BO82] the Hamiltonian function is interpreted as horizontal component of the resultant linear momentum on each cross–section. It has to be independent of x since no horizontal forces are acting on the fluid.

Additionally the system is reversible in the sense of Section 5.5. The reversibility operator is $\mathcal{R}(\Phi, z, \Psi, w) = (-\Phi, z, \Psi, -w)$. The reversibility of the system follows from $D\mathcal{R}\Omega_{can}D\mathcal{R}^* = -\Omega_{can}$ and the evenness of H in (Φ, w).

The trivial base flow $(\Phi, z, \Psi, w) = (x, 1, 1, 0)$ is a relative equilibrium in the sense of Section 6.7. Hence, to localize around it we let

$$(\Phi, z, \Psi, w) = (x, 1, 1, 0) + (u, \zeta, v, \xi), \quad \overline{H}(u, \zeta, v, \xi) = H(x + u, 1 + \zeta, 1 + v, \xi) - \int_0^1 v \, d\eta.$$

The additional term $-\int_0^1 v \, d\eta$ arises from the augmentation according to Section 6.7.

We find the following expansion

$$
\begin{aligned}
\overline{H}(u, \zeta, v, \xi) &= H_2(u, \zeta, v, \xi) + H_3(u, \zeta, v, \xi) + \mathcal{O}(\|(u, \zeta, v, \xi)\|^4) \\
H_2(u, \zeta, v, \xi) &= \tfrac{1}{2} \int_0^1 [(v - \zeta)^2 - u'^2] \, d\eta - \lambda \zeta^2 + \tfrac{1}{2b} \left(\xi + \int_0^1 \eta u' d\eta \right)^2 \\
H_3(u, \zeta, v, \xi) &= \tfrac{\zeta}{2} \int_0^1 [u'^2 - (v - \zeta)^2] \, d\eta + \tfrac{1}{b} \left(\xi + \int_0^1 \eta u' d\eta \right) \int_0^1 (v - \zeta) \eta u' d\eta.
\end{aligned}
\tag{9.2}
$$

Hence, the linear part of the Hamiltonian system is given by

$$
\frac{d}{dx}
\begin{pmatrix} u \\ \zeta \\ v \\ \xi \end{pmatrix}
= L
\begin{pmatrix} u \\ \zeta \\ v \\ \xi \end{pmatrix}
:=
\begin{pmatrix} 0 & 0 & I & 0 \\ 0 & 0 & 0 & 1 \\ -I & 0 & 0 & 0 \\ 0 & -1 & 0 & 0 \end{pmatrix}
\begin{pmatrix} \partial_u H_2 \\ \partial_\zeta H_2 \\ \partial_v H_2 \\ \partial_\xi H_2 \end{pmatrix}
=
\begin{pmatrix} v - \zeta \\ \tfrac{1}{b} \left(\xi + \int_0^1 \eta u' d\eta \right) \\ -u'' + \tfrac{1}{b} \left(\xi + \int_0^1 \eta u' d\eta \right) \\ (\lambda - 1)\zeta + \int_0^1 v \, d\eta \end{pmatrix}
$$

with the boundary conditions $u'(0) = 0$ and $u'(1) - \tfrac{1}{b} \left(\xi + \int_0^1 \eta u' d\eta \right) = 0$.

A direct calculation shows that $\sigma \in \mathbb{C}$ is in the spectrum of L if and only if $\sigma \cos \sigma = (\lambda - b\sigma^2) \sin \sigma$. As we are interested in the eigenvalues on the imaginary axis, we let $\sigma = is$, $s \in \mathbb{R}$, and obtain the relation $\tanh s = s/(\lambda + bs^2)$. For all parameters we have the eigenvalue $\sigma = 0$ which is due to the first integral $\int_0^1 v \, d\eta$. For $\lambda < 1$ there is a pair of solutions $s = \pm S(\lambda, b)$ with $S \to 0$ for $\lambda \to 1$ and $S \to \infty$ for $b \to 0$. For $\lambda > 1$ there is a curve $b = B(\lambda)$ with $B(1) = \tfrac{1}{3}$, $B' < 0$, and $B \to 0$ for $\lambda \to \infty$, such that for (λ, b) with $\lambda > 1$ and $b \in (0, B(\lambda))$ there are two pairs of eigenvalues on the imaginary axis. These two pairs collide when the parameters move through the curve and then move off the imaginary axis. For $\lambda > 1$ and $b > B(\lambda)$ only the trivial zero eigenvalue remains on the axis. See [AK89] and [IK91] for a more detailed analysis. We also refer to the first of these papers for the necessary functional analysis for proving the existence of a smooth and finite dimensional center manifold.

9.2   Small surface tension and λ close to 1

As one example we treat the case $\lambda \approx 1$ and $b < \frac{1}{3}$. It is one of the major remaining questions to find out whether for this parameter region small homoclinic solutions exist. They would correspond to solitary surface waves travelling with constant speed. For $b > \frac{1}{3}$ the bifurcations of these waves was shown in [AK89], and for $b = 0$ in [Mi88b]. In both cases the reduced system on the center manifold is a reversible two–dimensional system with an hyperbolic fixed point; thus the existence of homoclinic solutions follows easily without using the Hamiltonian structure.

In the case $b \in (0, \frac{1}{3})$ the situation changes as we have an additional pair of imaginary eigenvalues. For $\lambda = 1$ the eigenvalue $\sigma = 0$ is four–fold while the non–zero eigenvalues $\sigma = \pm is(b)$ are simple. After restricting ourselves to the submanifold $\int_0^1 v \, d\eta = 0$ we are led to a four–dimensional system having an equilibrium of saddle–center type for $\lambda < 1$, i.e. the linearization has a pair of purely imaginary eigenvalues and a pair of real eigenvalues, all being non–zero. Hence, the equilibrium point has one–dimensional stable and unstable manifolds and a two–dimensional center manifold. The question of existence of homoclinic orbits is thus equivalent to the question whether the unstable and stable manifold meet inside of the three–dimensional energy surface containing the equilibrium. For systems of this type a study of homoclinic solutions is carried through in [MHO91]. In certain cases very complicated bifurcation pictures with fractal nature arise when considering so–called n–homoclinic solutions. To decide whether this theory is applicable we have to calculate the first terms of the reduced Hamiltonian on the center manifold. If we are able to show that there is a compact flow–invariant region of the energy surface into which one part of the stable and one part of the unstable manifold are confined, then that theory would be applicable.

To start with, we construct the generalized eigenspaces:

$$
e_1 = \begin{pmatrix} 1 \\ 0 \\ 0 \\ 0 \end{pmatrix}, \;
e_2 = \begin{pmatrix} 0 \\ -1 \\ 0 \\ 0 \end{pmatrix}, \;
e_3 = \begin{pmatrix} -\frac{1}{2}\eta^2 \\ 0 \\ 0 \\ \frac{1}{3} - b \end{pmatrix}, \;
e_4 = \begin{pmatrix} 0 \\ \frac{1}{2} - b \\ \frac{1}{2} - b - \frac{1}{2}\eta^2 \\ 0 \end{pmatrix},
$$

where $Le_j = e_{j-1}$ for $j = 1, \ldots, 4$ with $e_0 = 0$, and

$$
e_5 = \begin{pmatrix} \cosh s\eta \\ 0 \\ 0 \\ 0 \end{pmatrix}, \;
e_6 = \begin{pmatrix} 0 \\ -\sinh s \\ s \cosh s\eta - \sinh s \\ 0 \end{pmatrix},
$$

where $s = s(b) > 0$, $Le_5 = -se_6$, and $Le_6 = se_5$. Moreover, using the canonical symplectic structure $\omega = \omega_{\text{can}}$ on $(L_2(0,1) \times I\!R)^2$ we find

$$\omega(e_1, e_4) = \omega(e_2, e_3) = \tfrac{1}{3} - b \quad \text{and}$$

$$\omega(e_5, e_6) = \alpha^2(b) := \tfrac{1}{4}(2s + \sinh(2s) + \tfrac{2}{s}(\cosh(2s) - 1)) > 0$$

while all the other combinations give zero.

Hence, the six–dimensional center manifold can be described in the form

$$(u, \zeta, v, \xi)^T = (\tfrac{1}{3} - b)^{-1/2}(q_1\, e_1 + q_2\, e_3 + p_1\, e_4 + p_2\, e_2)$$

$$+ \frac{1}{\alpha(b)}(q_3\, e_5 + p_3\, e_6) + h(q_1, \ldots, p_3),$$

where $h = \mathcal{O}(|(q,p)|^2)$ for $(q,p) \to 0$. According to Darboux's theorem in the form of Theorem 5.15 we can assume that (q,p) are already canonical coordinates. The reduced Hamiltonian $\widetilde{H}$ has the expansion

$$\widetilde{H}(q,p) = p_1 p_2 - \frac{1}{2}q_2^2 + \frac{1}{2s(b)}(q_3^2 + p_3^2) + \mathcal{O}(|(q,p)|^3 + (\lambda - 1)^2|(q,p)|^2).$$

From the action of $\mathcal{R}$ on the eigenfunctions e_j we see that the reduced reversibility operator on the center manifold is given by $\widetilde{R}(q,p) = (-q,p)$. According to Theorem 5.16 we know that $\widetilde{H} \circ \widetilde{R} = \widetilde{H}$, i.e. $\widetilde{H}$ is even in q. Of course $\widetilde{H}$ does not depend on the cyclic variable q_1. Hence, p_1 is the first integral corresponding to $\int_0^1 v\, d\eta$. Since in the beginning the mass flow was normalized to 1, we now can restrict our attention to the case $p_1 \equiv 0$.

After using a linear canonical change of coordinates with

$$(x_1, x_2) = (p_3, p_2) + \mathcal{O}(|\lambda - 1||p|), \quad (y_1, y_2) = (-q_3, -q_2) + \mathcal{O}(|\lambda - 1||q|)$$

we can write the four–dimensional Hamiltonian system in the form

$$\widehat{H}(x,y) = \frac{s(b)}{2}(y_1^2 + x_1^2) - \frac{1}{2}(y_2^2 + \mu(\lambda, b)x_2^2) - \nu(\lambda, b)x_2^3 + \mathcal{O}(|y|^4 + x_2^4 + |x_1|^3). \quad (9.3)$$

Now we have $\widehat{H}(x, -y) = \widehat{H}(x, y)$. The values μ and ν have the expansion $\mu(\lambda, b) = (\lambda - 1)/(\tfrac{1}{3} - b) + \mathcal{O}(|\lambda - 1|^2)$ and $\nu(\lambda, b) = \tfrac{1}{2}(\tfrac{1}{3} - b)^{-3/2} + \mathcal{O}(|\lambda - 1|)$. The latter coefficient can easily be calculated from H_3 in (9.2).

9.3   Homoclinic solutions

The existence question of solitary waves is now reduced to the question whether the reduced system given by (9.3) has homoclinic solutions, i.e. solutions tending to $(x,y) = 0$

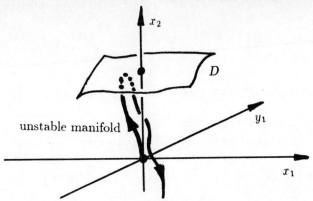

Figure 9.1: Flow on the energy surface $H = 0$.

for $t \to \pm\infty$. These solutions have to lie on the energy surface $\widehat{H}(x,y) = 0$. Since we are interested in the case $\lambda < 1$ we let $\mu = -\varepsilon^2$. We solve $\widehat{H} = 0$ for y_2 and obtain

$$y_2 = \pm\sqrt{s(b)(y_1^2 + x_1^2) + \varepsilon^2 x_2^2 - \nu(\varepsilon, b)x_2^3} + \mathcal{O}_\pm(x_1^2 + y_1^2 + |x_2|^3).$$

From $\dot{x}_2 = \frac{\partial}{\partial y_2}\widehat{H} = y_2 + \mathcal{O}(|(x,y)|^3)$ we find that y_2 is approximately the derivative of x_2. Thus we are able to visualize the flow in the (x_1, x_2, y_1)–space, keeping in mind that at each point there are two possible flow directions one having a positive and the other a negative $\dot{x}_2$-component. But a solution can only switch between these two vector fields when arriving at the surface $D = \{ (x,y) : \widehat{H}(x,y) = 0,\ y_2 = 0 \}$, see Figure 9.1. Because of $s(b)$ and ε being positive, $(x, y_1) = 0$ is an isolated point in D. Moreover, locally there is only one other part of D given by $x_2 = \varepsilon^2/\nu + \mathcal{O}(|x_1| + y_1^2)$. Hence, a solution starting below this surface will either go to negative x_2 values or hit, after infinite time, the origin.

The unstable manifold of the origin M_U is locally a smooth curve in (x, y)-space and has the approximation

$$(x_1, y_1) = \mathcal{O}(x_2^2), \quad y_2 = \varepsilon x_2 + \mathcal{O}(x_2^2).$$

It consists of two orbits, one starting with positive and increasing x_2. Along this solution x_2 will grow until the orbit hits D, then the x_2-component decreases. If the solution does not go back into the origin, it maintains the negative sign for $\dot{x}_2$ and, hence, has to leave a small neighborhood of size indenpent of ε. Similarly the other orbit of the unstable manifold starting with negative and decreasing x_2 has no possibility to return to the origin without leaving the small neighborhood.

As a conclusion we find, that bifurcating homoclinic solutions may exist, and if they exist they have a positive x_2-component which has a unique maximum of order ε^2/ν. Thus, no homoclinic doubling as described in [MHO91] occurs here.

Chapter 10

Necking of strips

We consider a two–dimensional strip composed out of a homogeneous isotropic material. There is a class of trivial solutions given by the homogeneous deformation $(u_1, u_2) = (\alpha y, \beta t)$, which satisfies zero traction conditions on the lateral boundaries if α is chosen as an appropriate function of β. These solutions are realized when the strip is exposed to a tensile load $\mathcal{F}_2$. It may happen that the function $\mathcal{F}_2 = \mathcal{F}_2(\beta)$ is non–monoton, and it is exactly then when *necking* occurs (cf. [Ow87]). This means that the strip will become narrower in a certain region without any additional loading. Below we give an analysis of this phenomenon.

10.1 The physical equations

The material behavior is described by a *stored–energy function* $\overline{W}$ which depends solely on the deformation gradient $F = \nabla u$. Frame indifference (objectivity) means that the material properties do not depend on ridig–body transformations, viz. $\overline{W}(RF) = \overline{W}(F)$ for all rotations R. The material is called isotropic when the material is also invariant under internal rotations, viz. $\overline{W}(FQ) = \overline{W}(F)$ (cf. [He88]). From frame indifference and isotropy it follows that $\overline{W}$ depends only on the singular values ν_1, ν_2 of F, which are defined to be the square roots of the eigenvalues of $F^T F$. Thus, we have $\overline{W}(F) = W(\nu_1(F), \nu_2(F))$, where W is symmetric in its arguments. The equilibrium equations for the strip $\Omega = \Sigma \times (-\ell, \ell)$ with $\Sigma = (-1/2, 1/2)$ are now given by minimizing the integral of the deformation energy over the whole body

$$\mathcal{I}(u) = \int_\Omega \overline{W}(\nabla u)\, dx + \int_\Sigma \varphi^- u(\cdot, -\ell)\, dy + \int_\Sigma \varphi^+ u(\cdot, \ell)\, dy. \tag{10.1}$$

Note that we have zero tractions on the lateral boundaries $\{\pm\frac{1}{2}\} \times (-\ell, \ell)$. In the following we will not specify the tractions $\varphi^\pm$ on the terminal boundaries, as we do not solve a particular boundary value problem; instead we classify certain classes of solutions existing for very long strips. The corresponding Euler–Lagrange equations are

$$\text{div}\left(\frac{\partial \overline{W}}{\partial F}(\nabla u)\right) = 0 \quad \text{in } \Omega,$$

$$\frac{\partial \overline{W}}{\partial F}(\nabla u)\begin{pmatrix} 1 \\ 0 \end{pmatrix} = \begin{pmatrix} 0 \\ 0 \end{pmatrix} \quad \text{for } y = \pm\tfrac{1}{2}. \tag{10.2}$$

Letting $Q = L_2(\Sigma, \mathbb{R}^2)$ the Lagrangian L is

$$L(u, \dot{u}) = \int_\Sigma \overline{W}(u', \dot{u})\, dy$$

where in the following $' = \partial/\partial y$ and $\dot{} = \partial/\partial t$. The associated Hamiltonian is defined in terms of u and the conjugate variable $v = \frac{\partial \overline{W}}{\partial F}(u', \dot{u})e_2$ which is exactly the stress distribution on the cross–section Σ. The strong ellipticity imposed below guarantees that this relation can be solved for

$$\dot{u} = m(u', v). \tag{10.3}$$

The Hamiltonian H is given by

$$H(u, v) = \int_\Sigma \left(v \cdot m(u', v) - \overline{W}(u', m(u', v))\right)\, dx. \tag{10.4}$$

As the strip has no internal loading and is stress–free on the lateral sides, the resultant force per cross–section

$$\mathcal{F} = \int_\Sigma v\, dy = \int_\Sigma \frac{\partial \overline{W}}{\partial F}(u', \dot{u})e_2\, dy \tag{10.5}$$

is constant along the solutions of (10.2); in fact it is the conjugate momentum to the cyclic variable $\int_\Sigma v\, dy$.

In the sequel we restrict our attention to solutions $u = (\alpha y, \beta t) + w$ with w_1 odd and w_2 even in y. Then $\mathcal{F}$ is of the form $\binom{0}{\mathcal{F}_2}$. For $\mathcal{F}_2 > 0$ the strip is under a tensile load while $\mathcal{F}_2 < 0$ corresponds to a compressive load.

First we consider homogeneous deformations $u = (\alpha y, \beta t)$, where we have $\nu_1 = \alpha$ and $\nu_2 = \beta$. The problem reduces to finding $\alpha = A(\beta)$ such that

$$\frac{\partial W}{\partial \nu_1}(\alpha, \beta) = 0. \tag{10.6}$$

The tensile force on the cross–section Σ is given by

$$\mathcal{F}_2(\beta) = \frac{\partial W}{\partial \nu_2}(A(\beta), \beta). \tag{10.7}$$

In the sequel we use the notation W_i, W_{ij} for the partial derivatives $\frac{\partial W}{\partial \nu_i}$, $\frac{\partial^2 W}{\partial \nu_i \partial \nu_j}$, $i, j = 1, 2$. Since

$$\frac{d\mathcal{F}_2}{d\beta} = W_{22} - \frac{W_{12}^2}{W_{11}}|_{(A(\beta),\beta)}, \tag{10.8}$$

non–monotonicity of $\mathcal{F}_2$ can only occur for non–convex functions W. According to [Ba77] non–convexity is a typical feature in good models for real elastic materials. The point is that strong ellipticity of (10.2) is still possible for non–convex W.

For these homogeneous deformations the deformation energy per cross–section and the Hamiltonian are given by

$$w(\beta) := W(A(\beta), \beta), \quad h(\beta) := H((A(\beta)y, \beta t)) = \beta W_2(A(\beta), \beta) - W(A(\beta), \beta). \tag{10.9}$$

Note that $W_1 = 0$ implies $\frac{dw}{d\beta} = \mathcal{F}_2$ and hence non–monotonicity of $\mathcal{F}_2$ is equivalent to non–convexity of w. We will be interested in β close to β_0 where $\mathcal{F}_2$ has a local maximum in β_0. Hence we assume $w''(\beta_0) = 0$ and $w'''(\beta_0) < 0$. For $h(\beta) = \beta w'(\beta) - w(\beta)$ this implies $h'(\beta_0) = 0$ and $h''(\beta_0) < 0$.

10.2 Strong ellipticity

Necessary and sufficient conditions on $W = W(\nu_1, \nu_2)$ for the strong ellipticity of (10.2) were first given in [KS77] for the two–dimensional case, and for three–dimensional elasticity in [SS83]. As we need many of the ideas developed there for our further analysis, we give a short, self–contained proof.

Lemma 10.1
Let $W = W(\nu_1, \nu_2) : \mathbb{R}^+ \times \mathbb{R}^+ \to \mathbb{R}$ being twice continuously differentiable and assume $\nabla_x u = \begin{pmatrix} \alpha & 0 \\ 0 & \beta \end{pmatrix} + \nabla_x w$, with $\nabla_x w$ small and $0 < \alpha < \beta$, then equation (10.2) is strongly elliptic if and only if the following inequalities are satisfied:

$$W_{11} > 0, \quad W_{22} > 0, \tag{10.10}$$

$$B := \frac{1}{\beta^2 - \alpha^2}(\beta W_2 - \alpha W_1) > 0, \tag{10.11}$$

$$\left| W_{12} + \frac{1}{\beta^2 - \alpha^2}(\alpha W_2 - \beta W_1) \right| < \sqrt{W_{11}W_{22}} + B. \tag{10.12}$$

(All the derivatives of W have to be evalutated in $(\nu_1, \nu_2) = (\alpha, \beta)$.)

Proof: We first expand the singular values ν_i in terms of $\nabla w = \begin{pmatrix} a & b \\ c & d \end{pmatrix}$. After some elementary calculations we arrive at

$$\nu_1 = \alpha + a + \frac{1}{2\alpha}\left(c^2 - \frac{(\alpha b + \beta c)^2}{\beta^2 - \alpha^2} \right) + \mathcal{O}(3), \quad \nu_2 = \beta + d + \frac{1}{2\beta}\left(b^2 + \frac{(\alpha b + \beta c)^2}{\beta^2 - \alpha^2} \right) + \mathcal{O}(3),$$

where $\mathcal{O}(3)$ represents terms of third order in ∇w. Inserting this into W yields

$$\overline{W}\left(\begin{pmatrix} \alpha+a & b \\ c & \beta+d \end{pmatrix}\right) = W(\alpha,\beta) + W_1(\alpha,\beta)a + W_2(\alpha,\beta)d + r_5bc + r_6ad$$
$$+\tfrac{1}{2}(r_1a^2 + r_2b^2 + r_3c^2 + r_4d^2) + \mathcal{O}(3), \tag{10.13}$$

where the coefficients r_i are given by

$$r_1 = W_{11}, \ r_2 = r_3 = B, \ r_4 = W_{22}, r_5 = \frac{1}{\beta^2 - \alpha^2}(\alpha W_2 - \beta W_1), \ \text{and } r_6 = W_{12}.$$

Hence the linearization of (10.6) in the form of (8.4) reads

$$\left.\begin{array}{r} r_1 w_1'' + r_2 \ddot{w}_1 + (r_5 + r_6)\dot{w}_2' = 0, \\ (r_5 + r_6)\dot{w}_1' + r_3 w_2'' + r_4 \ddot{w}_2 = 0, \end{array}\right\} \text{ for } (y,t) \in \Omega,$$
$$r_1 w_1' + r_6 \dot{w}_2 = r_5 \dot{w}_1 + r_3 w_2' = 0 \text{ for } y = \pm\tfrac{1}{2}. \tag{10.14}$$

According to (8.5) equation (10.14) is strongly elliptic if and only if

$$r_1\xi_1^2\eta_1^2 + r_2\xi_1^2\eta_2^2 + 2(r_5 + r_6)\xi_1\xi_2\eta_1\eta_2 + r_3\xi_2^2\eta_1^2 + r_4\xi_2^2\eta_2^2 > 0$$

for all ξ_i, η_i with $\xi_1^2 + \xi_2^2$, $\eta_1^2 + \eta_2^2 \neq 0$. However, this is the case if and only if

$$r_1, \ r_2, \ r_3, \ r_4 > 0, \ \text{and } |r_5 + r_6| < \sqrt{r_1 r_4} + \sqrt{r_2 r_3}$$

holds. Substituting the relations for r_i gives the result. $\square$

Note that (10.13) also shows that the quadratic part of the stored–energy function is non–negative if and only if

$$r_1 r_4 \geq r_6^2, \ r_2 r_3 \geq r_5^2. \tag{10.15}$$

If equality occurs only in one relation strong ellipticity is implied. If both inequalities are strict then $\overline{W}$ is locally convex; and hence local bifurcations are excluded. Above we have shown that non–monotonicity of $\mathcal{F}_2(\beta)$ occurs for $r_1 r_4 < r_6^2$.

Remark: For $\alpha \to \beta$ the coefficients B and r_5 converge to the limits

$$B = \frac{1}{2\beta}(\beta W_{11} + W_1 - \beta W_{12}), \ r_5 = \frac{1}{2\beta}(\beta W_{11} - W_1 - \beta W_{12}).$$

In particular, for an unstressed material (i.e. $\alpha = \beta = 1$) with Lamé constants λ and μ we have $r_1 = r_4 = \lambda + 2\mu$, $r_2 = r_3 = \mu$, $r_5 = \mu$, $r_6 = \lambda$, which is realized by functions of the form

$$W(\alpha,\beta) = \frac{1}{2}(\lambda + 2\mu)((\alpha - 1)^2 + (\beta - 1)^2) + \lambda(\alpha - 1)(\beta - 1) + \mathcal{O}(|\alpha - 1|^3 + |\beta - 1|^3).$$

Hence, strong ellipticity holds for $\mu > 0$, $|\lambda + \mu| < \lambda + 3\mu$. In view of the first inequality the second is equivalent to $\lambda + 2\mu > 0$. However according to (10.15) convexity holds only when $\lambda + \mu > 0$.

We now show that non–monotinicity of $\mathcal{F}_2$ is completely consistent with strong ellipticity, actually even with polyconvexity. Consider the function

$$W(\alpha, \beta) = a(\alpha) + a(\beta) + b(\alpha\beta),$$

where $a' \geq 0$ and a'', $b'' > 0$. According to [Ba77] W is polyconvex and hence strongly elliptic.

From (10.8) we obtain

$$\frac{d}{d\beta}\mathcal{F}_2(\beta) = a''(\beta) + \alpha^2 b''(\alpha\beta) - \frac{(\alpha\beta b''(\alpha\beta) + b'(\alpha\beta))^2}{a''(\alpha) + \beta^2 b''(\alpha\beta)}$$

Thus, we can make this expression negative by decreasing $a''(\beta)$ and $b''(\alpha\beta)$ while keeping and $|b'(\alpha\beta)|$ and $a''(\alpha)$ large.

We can construct an explicit example as follows. Since $(\alpha, \beta) = (1, 1)$ should be the natural state we require $a'(1) + b'(1) = 0$. Then the Lamé constants are given by $\mu = (a''(1) + a'(1))/2$ and $\lambda = b''(1) - a'(1)$. We want non–monotinicity to occur for $(\alpha, \beta) = (1/2, 4)$ and assume $a'(1) = -b'(1) = \mu$, $a'(1/2) = \mu/2$, and $b'(2) = -\mu/8 > b'(1)$. We are still free to choose $a''(1/2)$, $a''(4)$, and $b''(2)$ as arbitrary positive numbers. Then it is always possible to construct smooth a and b satisfying all aforementioned conditions. For instance, taking $b''(2) = \mu/64$ we obtain

$$\frac{d}{d\beta}\mathcal{F}_2(\beta) = a''(4) + \frac{\mu}{256}\left(1 - \frac{9\mu}{4a''(1/2) + \mu}\right)$$

For $a''(1/2) < 2\mu$ the last term is negative and thus a positive $a''(4)$ can be chosen such that the sum is still negative.

In particular, we are able to construct a family of stored–energy functions $W_\varepsilon = W_\varepsilon(\alpha, \beta)$ such that the corresponding reduced function $w_\varepsilon = w_\varepsilon(\beta)$ of (10.9) satisfies

$$w_\varepsilon(\beta) = w_0 + \mathcal{F}_2(\beta_0)(\beta - \beta_0) - \varepsilon(\beta - \beta_0)^2 + (\beta - \beta_0)^4 + \mathcal{O}(|\beta - \beta_0|^5 + \varepsilon^2|\beta - \beta_0|)$$

for $(\beta, \varepsilon) \to (\beta_0, 0)$ and some $\beta_0 > 1$.

10.3 The dynamical formulation

From now on we restrict our attention to the case were β is close to β_0 with $\frac{d}{d\beta}\mathcal{F}_2(\beta_0) = 0$. At $(\alpha, \beta) = (A(\beta_0), \beta_0)$ we assume the relations $W_1 = 0$, $W_2 > 0$, $W_{ij} > 0$, $W_{12}^2 =$

$W_{11}W_{22} > 0$, and $r_2 r_3 > r_5^2$. As a consquence all r_i are positive and strong ellipticity holds near $(A(\beta_0), \beta_0)$.

We write the nonlinear equilibrium equation (10.2) as a differential system which is emanable to the center manifold theorem 2.1. Therefore, we use the methods presented in [Mi88a] and in [Mi88c].

As in [Mi88a] we have to choose new unknowns in which the nonlinear boundary conditions $(10.2)_2$ reduce to linear ones. In particular, it is not possible to do the functional analysis in the canonical variables (u, v). From (10.13) we obtain

$$\begin{pmatrix} 0 \\ 0 \end{pmatrix} = \frac{\partial \overline{W}}{\partial F} \left(\begin{pmatrix} \alpha & 0 \\ 0 & \beta \end{pmatrix} + \nabla w \right) \begin{pmatrix} 1 \\ 0 \end{pmatrix} = \begin{pmatrix} r_1 w_1' + r_6 \dot{w}_2 \\ r_3 w_2' + r_5 \dot{w}_1 \end{pmatrix} + \mathcal{O}(2).$$

As r_1, r_3, r_5, and r_6 are positive, this equation gives locally a one–to–one correspondence between w' and $\dot{w}$, i.e. we have $w_i' = t_i(\dot{w})$. Now define $\widetilde{w} = (\widetilde{w}_1, \widetilde{w}_2)$ by $\widetilde{w}_1 = -\frac{r_3}{r_5} t_2(\dot{w})$, $\widetilde{w}_2 = -\frac{r_1}{r_6} t_1(\dot{w})$ to obtain, instead of $(10.2)_2$, the linear boundary conditions

$$r_1 w_1' + r_6 \widetilde{w}_2 = 0, \; r_3 w_2' + r_5 \widetilde{w}_1 = 0 \; \text{ for } y = \pm \frac{1}{2}. \tag{10.16}$$

Note also that $\dot{w}$ can be expressed by $\widetilde{w}$ through

$$\dot{w}_i = \widetilde{w}_i + n_i(\widetilde{w}) \quad \text{with } n_i = \mathcal{O}(|\widetilde{w}|^2), \; i = 1, 2. \tag{10.17}$$

Inserting (10.17) into the field equation $(10.2)_1$ we are able to solve for $\dot{\widetilde{w}}$ as r_2 and r_4 are positive:

$$\begin{aligned} \dot{\widetilde{w}}_1 &= -\frac{r_1}{r_2} w_1'' - \frac{r_5 + r_6}{r_2} \widetilde{w}_2' + n_3(w', w'', \widetilde{w}, \widetilde{w}'), \\ \dot{\widetilde{w}}_2 &= -\frac{r_3}{r_4} w_2'' - \frac{r_5 + r_6}{r_4} \widetilde{w}_1' + n_4(w', w'', \widetilde{w}, \widetilde{w}'). \end{aligned} \tag{10.18}$$

Altogether the nonlinear equation (10.2) can now be written in the form

$$\frac{d}{dt} \begin{pmatrix} w \\ \widetilde{w} \end{pmatrix} = \mathsf{L} \begin{pmatrix} w \\ \widetilde{w} \end{pmatrix} + n(w, \widetilde{w}), \tag{10.19}$$

where $(w, \widetilde{w})$ is considered as an element of the Hilbert space

$$X = \{ (w, \widetilde{w}) \in H^1(\Sigma, \mathbb{R}^2) \times H^0(\Sigma, \mathbb{R}^2) : w_1, \widetilde{w}_1 \text{ odd in } y, \, w_2, \widetilde{w}_2 \text{ even in } y \}.$$

The operator $\mathsf{L} : D(\mathsf{L}) \to X$ is given by

$$D(\mathsf{L}) = \{ (w, \widetilde{w}) \in X \cap H^2(\Sigma, \mathbb{R}^2) \times H^1(\Sigma, \mathbb{R}^2) : (w, \widetilde{w}) \text{ satisfies (10.16)} \},$$

$$\mathsf{L} \begin{pmatrix} w \\ \widetilde{w} \end{pmatrix} = \begin{pmatrix} \widetilde{w}_1 \\ \widetilde{w}_2 \\ -\frac{r_1}{r_2} w_1'' - \frac{r_5 + r_6}{r_2} \widetilde{w}_2' \\ -\frac{r_3}{r_4} w_2'' - \frac{r_5 + r_6}{r_4} \widetilde{w}_1' \end{pmatrix}$$

As the nonlinear mapping n is a pointwise function of the arguments indicated in (10.18), depending only linearly on w'' and $\widetilde{w}'$, we see that it generates a smooth map from $D(\mathsf{L})$ into X. (Use that $H^1(\Sigma)$ is continuously embedded into $C^0(\Sigma)$).

Lemma 10.2

The linear operator L has a discrete spectrum, and 0 is the only eigenvalue on the imaginary axis. The generalized eigenspace is spanned by the four vectors

$$\varphi_1 = \begin{pmatrix} 0 \\ 1 \\ 0 \\ 0 \end{pmatrix}, \quad \varphi_2 = \frac{1}{r_1}\begin{pmatrix} -r_6 y \\ 0 \\ 0 \\ r_1 \end{pmatrix}, \quad \varphi_3 = \frac{r_6}{2r_1 r_3}\begin{pmatrix} 0 \\ r_5 y^2 \\ -2r_3 y \\ 0 \end{pmatrix},$$

$$\varphi_4 = \frac{r_6}{24 r_1^2 r_3}\begin{pmatrix} 3(r_5^2 - r_2 r_3)y + 4(r_2 r_3 - r_5^2 - r_5 r_6)y^3 \\ 0 \\ 0 \\ 12 r_1 r_5 y^2 \end{pmatrix},$$

which satisfy $\mathsf{L}\varphi_1 = 0$ and $\mathsf{L}\varphi_{i+1} = \varphi_i$ for $i = 1, 2, 3$. Moreover, along the imaginary axis the resolvent satisfies the estimate

$$\|(\mathsf{L} - is)^{-1}\|_{X \to X} = \mathcal{O}(\frac{1}{|s|}) \quad \text{for } |s| \to \infty.$$

Proof: The existence of the resolvent can be proved as in [Mi88a]. Therefore we note that the strong ellipticity implies that the inhomogeneous problem (10.14) with right–hand side $(f_1, f_2) \in L_2(\widetilde{\Omega})$ and zero–traction data on $\partial\widetilde{\Omega}$ for bounded smooth $\widetilde{\Omega} \subset \mathbb{R}^2$ has a solution $u \in H^2(\widetilde{\Omega})$ (cf. [HM83]). Hence by relating the resolvent $(\mathsf{L} - is)^{-1}$ to an appropriate problem for functions $w : \Sigma \times \mathbb{R} \to \mathbb{R}^2$, which are periodic in t with period $2\pi/|s|$, the estimate follows as in [Mi88c].

Moreover the resolvent is a compact operator, implying that the spectrum is discrete with eigenvalues of finite multiplicity. In particular we are able to find the whole eigenspace to the eigenvalue zero (i.e. the generalized kernel) by solving successively $\mathsf{L}\varphi = \psi$ where ψ is already known to be in the generalized kernel. An elementary calculation now yields the generalized eigenvectors φ_i.

To show that there are no further eigenvalues on the imaginary axis it is sufficient to show that the homogeneous equation (10.14) has no nontrivial t–periodic solutions. To this end assume that w solves (10.14) and satisfies $w(x, t) = w(x, t + p)$ for some $p > 0$. Then multiply (10.14)$_1$ with w and integrate over $\Omega_p = \Sigma \times (0, p)$. Using the boundary conditions (10.14)$_2$ and the periodicity, integration by parts results in

$$0 = \int_{\Omega_p} \frac{1}{2}\left((\sqrt{r_1}w_1' + \sqrt{r_4}\dot{w}_2)^2 + +\frac{1}{r_2}(r_2\dot{w}_1 + r_5 w_2')^2 + \frac{r_2 r_3 - r_5^2}{r_2}w_2'^2 \right) dy\,dt.$$

Since $r_5^2 < r_2 r_3$, all terms in the integral are non–negative implying $\dot{w}_1 = w_2' = \sqrt{r_1}w_1' + \sqrt{r_4}\dot{w}_2 = 0$. Hence, the only possible solutions are constant. $\square$

This shows that Theorem 2.1 is applicable, and thus equation (10.19) has a center manifold which is four–dimensional and is spanned over the generalized kernel, i.e.

$$M_C = \{\, (w, \tilde{w}) \in D(\mathsf{L}) \;:\; (w, \tilde{w}) = \sum_{i=1}^{4} c_i \varphi_i + h(c_1, \ldots, c_4) \,\},$$

where $h = \mathcal{O}(|c|^2)$ for $c \to 0$. The reduced equation has the form

$$\frac{d}{dt}c = Ac + f(c), \quad \text{where } A = \begin{pmatrix} 0 & 1 & 0 & 0 \\ 0 & 0 & 1 & 0 \\ 0 & 0 & 0 & 1 \\ 0 & 0 & 0 & 0 \end{pmatrix} \text{ and } f = \mathcal{O}(|c|^2). \tag{10.20}$$

10.4   The reduced Hamiltonian problem

To find the Hamiltonian and whence the Lagrangian formulation of the reduced equation we first note that the problem is invariant under translations $u \to u + (0, \gamma)$ for every $\gamma \in \mathbb{R}$. Hence the mean value of u_2 is a cyclic variable, i.e. H does not depend on it, with the conjugate variable $\mathcal{F}_2 = \int_\Sigma v_2 dy$ which is a first integral of the system. It would be possible to reduce the whole system by facoring out $\int_\Sigma u_2 dy$ and fixing the value of $\mathcal{F}_2$. Instead of doing this we want to apply the theory of relative equilibra given in Section 6.7. Note that the solution $(u_0, v_0) = (A(\beta_0)y, \beta_0 t)$ is a relative equilibrium in this sense. The Lie group is equal to $\mathbb{R}$ with the usual addition. In this case the Lie algebra is $\mathbb{R}$ with the trivial Lie bracket $[\xi_1, \xi_2] = 0$ for all $\xi_i \in \mathbb{R}$.

We introduce the augmented Hamiltonian according to (6.13). Therefore let $w = u - u_0$, $\tilde{v} = v - v_0$, and

$$\begin{aligned} H_{aug}(w, \tilde{v}) &= H(u_0 + w, v_0 + \tilde{v}) - \beta_0 \mathcal{F}_2(u_0 + w, v_0 + \tilde{v}) + W(A(\beta_0), \beta_0), \\ \mathcal{F}(w, \tilde{v}) &= \int_\Sigma \tilde{v}_2 dy = \mathcal{F}_2 - W_2(A(\beta_0), \beta_0). \end{aligned} \tag{10.21}$$

The augmented Hamiltonian H_{aug} generates exactly the corresponding equations for $(w, \tilde{v})$ and satisfies the expansion

$$H_{aug}(w, \tilde{v}) = \int_\Sigma \frac{1}{2}\left(\frac{1}{r_2}(\tilde{v}_1 - r_5 w_2')^2 + \frac{1}{r_4}(\tilde{v}_2 - r_6 w_1')^2 - r_1 w_1'^2 - r_3 w_2'^2 \right) dx + \mathcal{O}(3).$$

Of course in the $(w, \tilde{v})$–coordinates we still have the canonical symplectic structure on $T^* L_2(\Sigma)^2$. To find the reduced structure on the center manifold M_C we have to transform it into the $(w, \tilde{v})$–coordinates. The vectors φ_j of Lemma 10.2 take, due to

$$\begin{pmatrix} \tilde{v}_1 \\ \tilde{v}_2 \end{pmatrix} = \begin{pmatrix} r_2\dot{w}_1 + r_5 w'_2 \\ r_4\dot{w}_2 + r_6 w'_1 \end{pmatrix} + \mathcal{O}(2) = \begin{pmatrix} r_2\tilde{w}_1 + r_5 w'_2 \\ r_4\tilde{w}_2 + r_6 w'_1 \end{pmatrix} + \mathcal{O}(2),$$

the form

$$\tilde{\varphi}_1 = \begin{pmatrix} 0 \\ 1 \\ 0 \\ 0 \end{pmatrix}, \quad \tilde{\varphi}_2 = \frac{1}{r_1} \begin{pmatrix} -r_6 y \\ 0 \\ 0 \\ 0 \end{pmatrix}, \quad \tilde{\varphi}_3 = \frac{r_6}{2r_1 r_3} \begin{pmatrix} 0 \\ r_5 y^2 \\ 2(r_5^2 - 2r_3)y \\ 0 \end{pmatrix},$$

$$\tilde{\varphi}_4 = \frac{r_6}{24 r_1^2 r_3} \begin{pmatrix} 3(r_5^2 - r_2 r_3)y + 4(r_2 r_3 - r_5^2 - r_5 r_6)y^3 \\ 0 \\ 0 \\ 3r_6(r_5^2 - r_2 r_3)(1 - 4y^2) \end{pmatrix}.$$

An elementary calculation results in $\Omega_{ij}^0 = \hat{\omega}_{can}(\tilde{\varphi}_i, \tilde{\varphi}_j)$ with

$$\Omega^0 = \nu \begin{pmatrix} 0 & 0 & 0 & -1 \\ 0 & 0 & 1 & 0 \\ 0 & -1 & 0 & -\kappa \\ 1 & 0 & \kappa & 0 \end{pmatrix} \quad \text{with } \nu = \frac{r_4(r_2 r_3 - r_5^2)}{12 r_1 r_3}, \quad \kappa = \frac{2 r_2 r_3 - 2 r_5^2 + r_5 r_6}{20 r_1 r_3}.$$

Defining $\psi_1 = \tilde{\varphi}_1$, $\psi_2 = \tilde{\varphi}_2$, $\psi_3 = \frac{1}{\nu}(\kappa \tilde{\varphi}_2 - \tilde{\varphi}_4)$, and $\psi_4 = \frac{1}{\nu}\tilde{\varphi}_3$, the center manifold can be written as

$$\begin{pmatrix} w \\ v \end{pmatrix} = \sum_{i=1}^{4} d_i \psi_i + \hat{h}(d_2, d_3, d_4), \tag{10.22}$$

where the reduced symplectic form in d–coordinates reads

$$\tilde{\Omega}(d) = \begin{pmatrix} 0 & 0 & 1 & 0 \\ 0 & 0 & 0 & 1 \\ -1 & 0 & 0 & 0 \\ 0 & -1 & 0 & 0 \end{pmatrix} + \mathcal{O}(|(d_2, d_3, d_4)|^2).$$

Inserting (10.22) into H_{aug} and $\mathcal{F}$ from (10.21) results, after a lengthy calculation, in

$$\begin{aligned} \tilde{H}(d_2, d_3, d_4) &= d_2 d_3 + \frac{\kappa}{2\nu} d_3^2 + \frac{1}{2\nu} d_4^2 + \mathcal{O}(|(d_2, d_3, d_4)|^3), \\ \tilde{\mathcal{F}}(d_2, d_3, d_4) &= d_3 + \mathcal{O}(|(d_2, d_3, d_4)|^2). \end{aligned} \tag{10.23}$$

It is a good control of the theory to get exactly these coefficients. Only they guarantee the equivalence of the linear part of the Hamiltonian system $\dot{d} = \tilde{J}(d) d\tilde{H}$ with that of (10.20).

Note that the coordinate d_1 does appear neither in $\tilde{\Omega}$, $\tilde{H}$, nor in $\tilde{\mathcal{F}}$ due to the invariance under translations. Using the results of Chapter 5 we know that it is possible to find canonical coordinates $(\bar{q}_1, \bar{q}_2, \bar{p}_1, \bar{p}_2) = (d_1, \dots, d_4) + \mathcal{O}(|(d_2, d_3, d_4)|^3)$. Denote the corresponding Hamiltonian by $\bar{H} = \bar{H}(\bar{q}_2, \bar{p}_1, \bar{p}_2)$. In these coordinates we may reverse the augmentation introduced above. For this purpose let $q_1 = \bar{q}_1 + \beta_0 t$, $q_2 = \bar{q}_2$, $p_1 = \mathcal{F}_2(\beta_0) + \bar{p}_1$, $p_2 = \bar{p}_2$, and

$$
\begin{aligned}
\tilde{H}(q_2, p_1, p_2) = {} & -W(A(\beta_0), \beta_0) + \beta_0 p_1 + q_2(p_1 - f_0) + \frac{\kappa}{2\nu}(p_1 - f_0)^2 + \frac{1}{2\nu}p_2^2 \\
& + a_0(p_1 - f_0)^3 + a_1(p_1 - f_0)^2 q_2 + a_2(p_1 - f_0)q_2^2 + a_3 q_2^3 \\
& + \mathcal{O}(q_2^4 + (p_1 - f_0)^4 + |p_2|^3).
\end{aligned}
$$

We use the abbreviation $f_0 = \mathcal{F}_2(\beta_0)$ and have introduced the additional coefficients $a_0, \ldots, a_3$ which could be calculated from the material law $W = W(\alpha, \beta)$. However, we skip this computation and rather assume that a_3 is positive. Below it is shown that this is equivalent to the assumption that $\mathcal{F}_2(\beta)$ has a local maximum in β_0, or more precisely $\frac{d^2}{d\beta^2}\mathcal{F}_2(\beta_0) < 0$.

This Hamiltonian system can be transfered via $\dot{q}_1 = \beta_0 + q_2 + \frac{\kappa}{\nu}(p_1 - f_0) + \mathcal{O}(2)$ and $\dot{q}_2 = \frac{1}{\nu}p_2 + \mathcal{O}(2)$ into the Lagrangian system generated by

$$
\begin{aligned}
\tilde{L}(q_2, \dot{q}_1, \dot{q}_2) = {} & W(A(\beta_0), \beta_0) + f_0(\dot{q}_1 - \beta_0) + \frac{\nu}{2\kappa}\delta^2 + \frac{\nu}{2}\dot{q}_2^2 - a_0(\frac{\nu}{\kappa}\delta)^3 \\
& - a_1(\frac{\nu}{\kappa}\delta)^2 q_2 - a_2(\frac{\nu}{\kappa}\delta)q_2^2 - a_3 q_2^3 + \mathcal{O}(q_2^4 + (\dot{q}_1 - \beta_0)^4 + |\dot{q}_2|^3),
\end{aligned}
$$

where $\delta = \dot{q}_1 - \beta_0 - q_2$. The reduced Lagrangian $\tilde{L}$ is locally very similar to the one derived in [Ow87] by the projection method. The analysis done there can be applied to our case analogously.

To see how the necking solutions arise we go back to the Hamiltonian system in (q, p)–coordinates. For tensile loadings just below the critical loading $f_0 = \mathcal{F}_2(\beta_0)$ we show that there exist localized necks as well as periodic ones. We fix a small $\varepsilon > 0$ and let $p_1 = f_0 - \varepsilon^2 = \text{const}$. The remaining Hamiltonian system for (q_2, p_2) is given by

$$
\dot{q}_2 = \frac{\partial}{\partial p_2}\tilde{H}(q_2, f_0 - \varepsilon^2, p_2) = \frac{1}{\nu}p_2 + \mathcal{O}(\varepsilon^3 + |q_2|^3 + p_2^2),
$$

$$
\dot{p}_2 = -\frac{\partial}{\partial q_2}\tilde{H}(q_2, f_0 - \varepsilon^2, p_2) = \varepsilon^2 - 3a_3 q_2^2 + \mathcal{O}(\varepsilon^3 + |q_2|^3 + p_2^2).
$$

There are two families of t–independent solutions $(q_2, p_2) = (Q^\pm(\varepsilon), 0)$ with $Q^\pm(\varepsilon) = \pm\varepsilon/\sqrt{3a_3} + \mathcal{O}(\varepsilon^2)$. They correspond to the homogeneous solutions $u = (A(\beta)y, \beta t)$ of the original equations. Since the dependence of $\mathcal{F}_2$ on β is non–monoton there are two solutions for $p_1 < f_0$ and no solution for $p_1 > f_0$. This justifies the previous assumption $a_3 > 0$. The solution with negative q_2 is a saddle and the other is a center. The full phase portrait is shown in the Figure 10.1.

Moreover, there is a solution $(q_2, p_2) = (Q_{\text{hom}}^\varepsilon(\cdot), P_{\text{hom}}^\varepsilon(\cdot))$ homoclinic to the saddle $Q^-(\varepsilon)$. It has the expansion

$$
Q_{\text{hom}}^\varepsilon = \frac{\varepsilon}{\sqrt{3a_3}}\left(\frac{6}{1 + \cosh(\gamma t)} - 1\right) + \mathcal{O}(\varepsilon^2)
$$

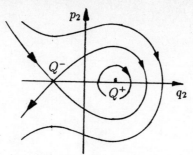

Figure 10.1: Phase portrait

where $\gamma^2 = \frac{2\varepsilon}{\nu}\sqrt{3a_3}$. Since the eigenfunction $\tilde{\varphi}_2$ is such that large q_2 corresponds to thin portions of the strip, this homoclinic solution describes a localized neck in an infinite strip. Similarly, the periodic solutions inside of the homoclinic loop can be interpreted as periodic necks.

The projection method in [Ow87] looks for solutions satisfying the constraint $(u_1, u_2) = ((\alpha_0 - q_2(t))y, q_1(t))$, which is motivated by the assumption that shear forces do not play an important role in the necking phenomenon. The projected energy density is then defined as

$$W_P(q_2, \dot{q}_1, \dot{q}_2) = \frac{1}{2}\int_{-1}^{1} \overline{W}\left(\begin{pmatrix} \alpha_0 - q_2 & -\dot{q}_2 y \\ 0 & \dot{q}_1 \end{pmatrix}\right) dy.$$

Using (10.13) and the fact $\dot{q}_1 \approx \beta_0$ we are led to the expansion

$$W_P(q_2, \dot{q}_1, \dot{q}_2) = W(\alpha_0, \beta_0) + f_0(\dot{q}_1 - \beta_0) + \frac{r_4}{2}\delta^2 + \frac{r_2 r_4}{6r_1}\dot{q}_2^2 + \mathcal{O}(3).$$

Comparing with the exact Lagrangian $\tilde{L}$ we find that both coefficients of the quadratic terms are different. We conclude that shear effects are relevant in the necking problem, at least in the transition region where the thickness changes appear.

Chapter 11

Saint–Venant's problem

Most of the theory developed above originated from the questions related to Saint–Venant's problem which will be treated now. It is concerned with the connection between the three–dimensional beam theory and the one–dimensional rod theory.

It is proved in [Mi88c] that all equilibrium deformation of an infinite beam without external forces and having only small bounded strains are actually contained in a finite–dimensional center manifold. Moreover, the invariance of the problem under rigid–body transformations was exploited to show that the flow on the center manifold is described by a differential equation having exactly the form of the classical rod equations of Kirchhoff [Ki59] and Antman [An72]. This reduction process gives a mathematically rigorous way for deriving the corresponding (nonlinear) material law of the rod from the material law of the material constituting the three–dimensional beam.

The question, left unanswered in previous work, is whether starting with a hyperelastic three–dimensional material leads necessarily to a hyperelastic rod. Hyperelasticity just means that the equilibrium equations are found as the Euler–Lagrange equations with respect to some deformation energy density, usually called the stored–energy function. In the context of the present work this question is exactly the one of reducing a Lagrangian system onto its center manifold.

We derive the associated reduced stored–energy functional for the rod up to terms of third order. On this way we encounter the so–called Saint–Venant solutions of linearized elasticity which play a basic role in constructing the center space for the beam problem. Thus, we are able to avoid the deficiencies of classical projection methods which were not able to find correct values for the torsional rigidity (cf. the discussion at the end of Section 11.3). Moreover, we are able to show that, up to this order, the system allows for natural reduction. This provides the satisfactory result that the projection method, when applied with care, gives the correct result.

The resulting rod equations will be studied briefly in Section 11.5, for a complete discussion we refer to [HM88]. We mainly emphasize how the classical reduction via symmetry can be applied to the reduced rod model obtained via the center manifold approach. In the general case the reduced phase space (see Section 5.1) is four–dimensional, and contains chaotic dynamics.

Cross–sectional symmetries and reflection symmetry with respect to a cross–section (reversibility) are studied in the last section.

11.1 The physical equations

Let $\Omega = \Omega_l = \Sigma \times (-l,l)$ be the undeformed prismatic body, where the cross–section Σ with coordinates $y = (y_1, y_2)$ is a bounded region in $\mathbb{R}^2$ with C^2–boundary $\partial\Sigma$ and where $(-l,l)$ is the domain of the axial coordinate t. For a deformation $\varphi : \Omega_l \to \mathbb{R}^3$ the symmetric *strain tensor* $E = \mathcal{E}(\nabla\varphi)$ is given by $\frac{1}{2}(\nabla_x\varphi^*\nabla_x\varphi - I)$, where x stands for (y_1, y_2, t). For *hyperelastic materials*, the material behavior is characterized by the stored–energy function $W = W(y, t, \nabla_x\varphi)$. By the invariance of the material properties under rigid–body transformations (also called frame indifference or objectivity) we know that W has to satisfy the relation $W(x, RF) = W(x, F)$ for every orthogonal matrix R. This implies that W is actually only a function of E, viz. $W(x, F) = \widetilde{W}(x, E)$. For simplicity, we assume that the material is homogeneous in the axial direction, i.e. W does not depend on t, and that the expansion

$$\widetilde{W}(y, E) = \frac{\lambda}{2}(\operatorname{tr} E)^2 + \mu \operatorname{tr}(E^2) + \mathcal{O}(|E|^3) \qquad \text{for } E \to 0, \tag{11.1}$$

holds, where the Lamé constants λ and μ are positive and y–independent. By rescaling we may assume $\mu = 1$ to simplify subsequent formulae.

The equilibrium equations for a beam loaded only at its terminal surfaces $\Sigma \times \{\pm l\}$ are the Euler–Lagrange equations for the functional

$$I(\varphi) = \int_{\Omega_l} W(y, \nabla_x\varphi)\, dx + \int_\Sigma (\varphi(l, \cdot)g_+ - \varphi(-l, \cdot)g_-)\, dy.$$

The Cauchy *stress tensor* $T(y, \nabla_x\varphi)$ is given by $\frac{\partial}{\partial F}W$, i.e. $T_{ij} = \frac{\partial}{\partial F_{ij}}W$ and has the form $T(y, F) = FS(y, E)$, where the symmetric Piola–Kirchhoff tensor S is given by $\frac{\partial}{\partial E}\widetilde{W}$. The Euler–Lagrange equations are

$$\operatorname{div}(T(y, \nabla_x\varphi)) = 0 \quad \text{in} \quad \Omega_l, \tag{11.2}$$

$$T(y, \nabla_x\varphi)n = 0 \quad \text{on} \quad \partial\Sigma \times (-l, l), \tag{11.3}$$

$$T(y, \nabla_x\varphi)n = g_\pm \quad \text{for} \quad (y, t) \in \Sigma \times \{\pm\}. \tag{11.4}$$

Equation (11.2) is strongly elliptic for small E due to our assumption $\lambda, \mu > 0$. In fact, strong ellipticity is a reasonable global assumptions on W (cf. [Ba77]).

Hence, the fiber derivative

$$\psi = \frac{\partial W}{\partial \dot{\varphi}} = \frac{\partial W}{\partial F}(\nabla\varphi, \dot{\varphi})e_3 = \begin{pmatrix} \phi_{3,1} + \dot{\phi}_1 \\ \phi_{3,2} + \dot{\phi}_2 \\ \lambda(\phi_{1,1} + \phi_{2,2}) + \lambda_2\dot{\phi}_3 \end{pmatrix} + \text{h.o.t.}$$

$(\nabla = \nabla_y)$ is locally invertible with inverse

$$\dot{\varphi} = r(\nabla\varphi, \psi) = \begin{pmatrix} \psi_1 - \phi_{3,1} \\ \psi_2 - \phi_{3,2} \\ \lambda_2^{-1}(\psi_3 - \lambda(\phi_{1,1} + \phi_{2,2})) \end{pmatrix} + \text{h.o.t.} \qquad (11.5)$$

Here we used the abbreviations $\phi_{i,j} = \partial\phi_i/\partial y_j$ and $\lambda_n = \lambda + n$. The Lagrangian L, the energy E, and the Hamiltonian H take the form

$$L(\varphi, \dot{\varphi}) = \int_\Sigma W(\nabla\varphi, \dot{\varphi})\, dy,$$

$$E(\varphi, \dot{\varphi}) = \int_\Sigma [\dot{\varphi} \cdot \frac{\partial W}{\partial F}(\nabla\varphi, \dot{\varphi})e_3 - W(\nabla\varphi, \dot{\varphi})]dy,$$

$$H(\varphi, \psi) = \int_\Sigma [r(\nabla\varphi, \psi) \cdot \psi - W(\nabla\varphi, r(\nabla\varphi, \psi))]dy.$$

The existence of the energy E as a first integral of (11.2) and (11.3) was already found in [Er77], but up to now the interpretation as a Hamiltonian was not worked out.

11.2 Symmetry under rigid–body transformations

The basic feature in Saint–Venant's problem is the invariance under the rigid–body transformations. This is inherited in the special form of W given by $W(F) = \widetilde{W}(E)$. Hence, neglecting the boundary conditions at the terminal surfaces in (11.4), any solution φ transfers into another one by $\widetilde{\varphi} = R\varphi + r$ where $(R, r) \in SO(3) \rhd I\!\!R^3$ represents the *rigid–body transformation*.

To recover the notations of Chapter 5 we call $G = SO(3) \rhd I\!\!R^3$ the Lie group of Euclidian transformations in $I\!\!R^3$. It acts on $\mathcal{Q} = Q = L_2(\Sigma, I\!\!R^3)$ by

$$\Phi_{(R,r)}(\varphi)(y) = R\varphi(y) + r, \quad \text{for } y \in \Sigma.$$

The trivial solution, at which the center manifold will be constructed, is the undeformed configuration of the beam given by the curve $\varphi_0(y, t) = (y, t)^T$. Thus we, are in the case of a relative equilibrium as discussed in Section 6.7. As a base point we choose $q_0 = \varphi_0(\cdot, 0) = (y, 0)^T$. The orbit $O_G(q_0) \subset \mathcal{Q}$ is a six–dimensional submanifold containing the

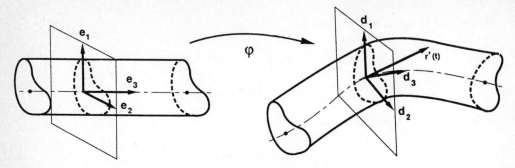

Figure 11.1: The rod as a constraint beam $(d_j = R(t)e_j)$.

whole curve φ_0. Particularly, the action is regular close to q_0. Obviously, the Lagrangian L and the Hamiltonian H are G–invariant. The slice construction for splitting Q into $G \times \tilde{Q}$ according to Theorem 5.1 was already done explicitly in [Mi88c]. The curve $g_0(t) = (R_0(t), r_0(t))$ in the Lie group G is given by $(\dot{R}_0, \dot{r}_0) = (0, e_3) = \xi_0$.

Because of the invariance of the system we obtain conserved quantities due to Noether's theorem. In our case they are well–known, namely the resultant force $\mathcal{F}$ and the resultant moment $\mathcal{M}$ per cross–section:

$$\mathcal{F} = \int_\Sigma \frac{\partial W}{\partial F}(\nabla_x\varphi)e_3 dy = \int_\Sigma \psi \, dy, \quad \mathcal{M} = \int_\Sigma \varphi \times \frac{\partial W}{\partial F}(\nabla_x\varphi)e_3 dy = \int_\Sigma \varphi \times \psi \, dy.$$

Traditionally the variational problem of beam theory is reduced, via the projection method, to a rod model by using the constraint ansatz

$$\varphi(y,t) = r(t) + R(t)\begin{pmatrix} y \\ 0 \end{pmatrix}, \quad \dot{\varphi}(y,t) = \dot{r}(t) + \dot{R}(t)\begin{pmatrix} y \\ 0 \end{pmatrix}.$$

(See Figure 11.1).

The projected energy functional is then defined by

$$W_P(r, R, \dot{r}, \dot{R}) = \int_\Sigma W\left(R\begin{pmatrix} 1 & 0 \\ 0 & 1 \\ 0 & 0 \end{pmatrix}, \dot{r} + \dot{R}\begin{pmatrix} y \\ 0 \end{pmatrix} \right) dy.$$

The invariance under rigid–body transformations implies that W_P has to have the form $W_P(r, R, \dot{r}, \dot{R}) = \widetilde{W}_P(R^T\dot{r}, R^T\dot{R})$ which is qualitatively exactly the same as that of the correct reduced Lagrangian W_{rod} derived below.

However, as is known since Saint–Venant in 1855 [dS55], such a method can not deliver the correct linearized torsional rigidity τ of the (unconstrained) beam. In fact, to calculate τ one has to solve a partial differential equation for the warping function Φ on the cross–section Σ, see Eqn. (11.7). This deficiency of the projection method comes from the constraint that the cross–section has to stay plane, which is not true because of warping. Our center manifold reduction will take care of this problem automatically.

11.3   The reduced Lagrangian system

Since we are in the case of a relative equilibrium we define $\phi(y,t) = \varphi(y,t) + (y,t)$ and the augmented Hamiltonian $H_{aug}(\phi,\psi) = H(\phi + (y,t),\psi) - \int_\Sigma \psi_3 dy$. Using (11.5) we do the expansion $H_{aug} = H_2 + \mathcal{O}(|(\nabla\phi,\psi)|^2)$ and arrive at

$$H_2(\phi,\psi) = \int_\Sigma \Big[\tfrac{1}{2}(\psi_1^2 + \psi_2^2 + \lambda_2^{-1}\psi_3^2) - \psi_1\phi_{3,1} - \psi_2\phi_{3,2} - \lambda\lambda_2^{-1}\psi_3(\phi_{1,1} + \phi_{2,2})$$
$$-\phi_{1,1}^2 - \phi_{2,2}^2 - \lambda\lambda_2^{-1}(\phi_{1,1} + \phi_{2,2})^2 - \tfrac{1}{2}(\phi_{1,2} + \phi_{2,1})^2 \Big]\, dy.$$

The associated linear operator $K : D(K) \to H^1(\Sigma)^3 \times L_2(\Sigma)^3$ with $D(K) = \{ (\phi,\psi) \in H^2(\Sigma)^3 \times H^1(\Sigma)^3 : B\binom{\phi}{\psi} = 0 \}$ is defined by

$$K\binom{\phi}{\psi} = \begin{pmatrix} \psi_1 - \phi_{3,1} \\ \psi_2 - \phi_{3,2} \\ \lambda_2^{-1}(\psi_3 - \lambda(\phi_{1,1} + \phi_{2,2})) \\ -\lambda_2^{-1}(4\lambda_1\phi_{1,11} + \lambda_2\phi_{1,22} + (3\lambda+2)\phi_{2,12} + \lambda\psi_{3,1}) \\ -\lambda_2^{-1}((3\lambda+2)\phi_{1,12} + \lambda_2\phi_{2,11} + 4\lambda\phi_{2,22} + \lambda\psi_{3,2}) \\ -\psi_{1,1} - \psi_{2,2} \end{pmatrix},$$

$$B\binom{\phi}{\psi} = \begin{pmatrix} 2\phi_{1,1}+\lambda\lambda_2^{-1}(2\phi_{1,1}+2\phi_{2,2}+\psi_3) & \phi_{1,2} + \phi_{2,1} \\ \phi_{1,2} + \phi_{2,1} & 2\phi_{2,2}+\lambda\lambda_2^{-1}(2\phi_{1,1}+2\phi_{2,2}+\psi_3) \\ \psi_1 & \psi_2 \end{pmatrix}\binom{n_1}{n_2},$$

where (n_1, n_2) is the unit outward normal vector at $\partial\Sigma$ in $\mathbb{R}^2$.

It is easily seen that K has a nontrivial kernel. In fact, the six infinitesimal rigid–body transformations (corresponding to the Lie algebra of the Euclidian group) give rise to the following vectors

$$x_1 = e_1,\ x_2 = e_2,\ x_3 = e_3,\ x_4 = -y_2e_1 + y_1e_2,\ x_5 = -y_1e_3,\ x_6 = -y_2e_3,$$

which satisfy $Kx_j = 0$ for $i = 1,\ldots,4$ and $Kx_i = x_{i-4}$ for $i = 5,6$. Hence, these vectors belong to the generalized kernel. However, since they form an isotropic subspace, the linear theory in Chapter 3 tells us that the symplectic generalized kernel $V(0)$ is at least 12–dimensional. Additional six vectors are found by inspecting the so–called Saint–Venant solutions of linear elasticity (see [dS55, Mi88c]). They are solutions polynomial in the axial variable and having non–zero resultants $\mathcal{F}$ and $\mathcal{M}$. We find

$$z_1 = \frac{1}{4\lambda_1}\begin{pmatrix} 0 \\ 0 \\ \Psi_1 \\ \Psi_{1,1} + \lambda(y_1^2 - y_2^2) \\ \Psi_{1,2} + 2\lambda y_1 y_2 \\ 0 \end{pmatrix},\ z_2 = \frac{1}{4\lambda_1}\begin{pmatrix} 0 \\ 0 \\ \Psi_2 \\ \Psi_{2,1} + 2\lambda y_1 y_2 \\ \Psi_{2,2} + \lambda(y_2^2 - y_1^2) \\ 0 \end{pmatrix},\ z_3 = \frac{1}{2\lambda_1}\begin{pmatrix} -\lambda y_1 \\ -\lambda y_2 \\ 0 \\ 0 \\ 0 \\ 2(3\lambda+2) \end{pmatrix},$$

$$z_4 = \begin{pmatrix} 0 \\ 0 \\ \Phi \\ \Phi_{,1} - y_2 \\ \Phi_{,2} + y_1 \\ 0 \end{pmatrix}, \quad z_5 = \frac{1}{4\lambda_1} \begin{pmatrix} \lambda(y_1^2 - y_2^2) \\ 2\lambda y_1 y_2 \\ 0 \\ 0 \\ 0 \\ -4(3\lambda+2)y_1 \end{pmatrix}, \quad z_6 = \frac{1}{4\lambda_1} \begin{pmatrix} 2\lambda y_1 y_2 \\ \lambda(y_2^2 - y_1^2) \\ 0 \\ 0 \\ 0 \\ -4(3\lambda+2)y_2 \end{pmatrix},$$

where Φ is the warping function and Ψ_1, Ψ_2 the flexure functions which satisfy the equations $\Delta\Phi = 0$, $\Delta\Psi_j = 8\lambda_1 y_j$. The vectors are numbered such that $Kz_j = x_j$ for $j = 3, \ldots, 6$ and $Kz_i = z_{i+4}$ for $i = 1, 2$. Thus, we have two Jordan chains of length 2 and two of length 4. The vectors z_1, z_2 correspond to flexure (or shear) solutions, z_3 to extension, z_4 to torsion, and z_5, z_6 to bending solutions.

Using the classical linear Saint–Venant principle (cf. [OY83, Mi90]) it is shown in [Mi88c] that the span of these twelve vectors is the whole center space and that a 12–dimensional center manifold can be constructed with the help of Theorem 2.1. Moreover, $\dim(G) = 6$ and Theorem 5.7 imply that the flow on the center manifold can be identified with a Hamiltonian system on T^*G equipped with its canonical symplectic form. It remains to be shown that the flow on the center manifold is a Lagrangian flow. Therefore we use Theorem 6.12. First we note that the condition (6.11) for C_{ξ_0} is satisfied according to Example 2 in Section 6.7. Second the kernel of K is four–dimensional, thus Theorem 6.5 guarantees that the flow on the center manifold is a Lagrangian flow. According to Theorem 6.9 the reduced Lagrangian problem can be posed on TG with an G–invariant $\widetilde{L}(r, R, \dot{r}, \dot{R})$, i.e. $\widetilde{L}(r, R, \dot{r}, \dot{R}) = W_{\text{rod}}(R^T\dot{r}, R^T\dot{R})$ with $W_{\text{rod}}(v, U) = \widetilde{L}((0, I), (v, U))$.

For the beam deformations this means that the variables $(R(t), r(t)) \in G = SO(3) \triangleright \mathbb{R}^3$ can be interpreted as the mean rotation and the mean translation of the cross–section $\Sigma \times \{t\}$ as in the constraint rod model of Figure 11.1. However, now the cross–sections do not stay planar. The reduced G–invariant Lagrangian $\widetilde{L}$ depends only on the values $(v, U) \in \mathbf{g}$. These are the well–known local strains

$$v = R^T\dot{r}, \qquad U = \begin{pmatrix} 0 & u_3 & -u_2 \\ -u_3 & 0 & u_1 \\ u_2 & -u_1 & 0 \end{pmatrix} = R^T\dot{R}$$

of flexure or shear (v_1, v_2), extension (v_3), bending (u_1, u_2), and torsion (u_3), see e.g. [An72, HM88]). Subsequently we identify the skew–symmetric matrix U with the vector u; then the commutator $U_1 U_2 - U_2 U_1$ can be identified with the vector product $u_1 \times u_2$. Hence, the rod equations are the Euler–Lagrange equations to the functional

$$\mathcal{I}(r, R) = \int_{-l}^{l} W_{\text{rod}}(R^T\dot{r}, R^T\dot{R}) \, dt,$$

In this sense the rod equations derived from a hyperelastic beam are again *hyperelastic*.

Yet, the reduced energy density W_{rod} is not unique, as the definition of the strains (v, u) is not due to possible G–invariant canonical coordinate changes in T^*G.

11.4   On natural reduction

We return to the linearized problem and show that the coordinates on the center space can be chosen such that the reduced and the projected Lagrangian coincide up to terms of third order. To simplify the analysis, we restrict ourselves to the case that the cross–section Σ is symmetric with respect to both y_i–axes. This implies

$$\Phi(a_1 y_1, a_2 y_2) = a_1 a_2 \Phi(y_1, y_2), \ \ \Psi_i(a_1 y_1, a_2 y_2) = a_i \Psi_i(y_1, y_2) \ \text{ for } (a_1, a_2) \in \{\pm 1\}^2. \ \ (11.6)$$

In (ϕ, ψ)–coordinates we have the canonical symplectic form $\omega = \omega_{\text{can}}$ and the vectors x_i, z_i satisfy

$$\omega(x_i, x_j) = 0, \quad \omega(x_i, z_j) = \delta_{ij} d_i \text{ for } i, j = 1, \ldots, 6;$$

$$\omega(z_k, z_{k+4}) = \kappa_k \text{ for } k = 1, 2; \quad \omega(z_i, z_j) = 0 \text{ if } |i - j| \neq 4.$$

Here we have made extensive use of the symmetry properties (11.6). Moreover, we have

$$d_5 = -d_1 = (3\lambda + 2)\lambda_1^{-1}\sigma_1, \ \ d_6 = -d_2 = (3\lambda + 2)\lambda_1^{-1}\sigma_2,$$

$$d_3 = (3\lambda + 2)\lambda_1^{-1}\sigma_0, \qquad \tau = d_4 = \int_\Sigma \{(\Phi_{,1} + y_2)^2 + (\Phi_2 - y_1)^2\} \, dy; \tag{11.7}$$

where $\sigma_0 = \int_\Sigma 1 \, dy$ and $\sigma_k = \int_\Sigma y_k^2 \, dy, \ k = 1, 2$. The relation $d_5 = -d_1$ is a consequence of the skew–symmetry of K with respect to ω (cf. (3.2)). We have $d_5 = \omega(x_5, z_5) = \omega(x_5, Kz_1) = -\omega(Kx_5, z_1) = -\omega(x_1, z_1) = -d_1$.

We choose a symplectic basis by keeping the x_i fixed and let

$$\overline{z}_k = \tfrac{1}{d_k} z_k + \alpha_k x_{k+4}, \ \ \overline{z}_{k+2} = \tfrac{1}{d_{k+2}} z_{k+2}, \ \ \overline{z}_{k+4} = \tfrac{1}{d_{k+4}} z_{k+4} + \alpha_{k+4} x_k, \ \ k = 1, 2.$$

Here, the α_i are real parameters, and $\{x_i, \overline{z}_j\}$ forms a symplectic basis if and only if

$$\kappa_1 + (\alpha_5 - \alpha_1)d_1^2 = \kappa_2 + (\alpha_6 - \alpha_2)d_2^2 = 0. \tag{11.8}$$

The reduced Hamiltonian can be found by calculating the reduced linear operator K_1. From

$$K \sum_{i=1}^6 (q_i x_i + p_i \overline{z}_i) = q_5 x_1 + q_6 x_2 + p_1(-\overline{z}_5 + (\alpha_1 + \alpha_5)x_1)$$

$$+ p_2(-\overline{z}_6 + (\alpha_2 + \alpha_6)x_2) + \sum_{i=3}^4 p_i x_i / d_i,$$

we obtain

$$K_1 = \begin{pmatrix} -C^* & \overline{D} \\ 0 & C \end{pmatrix} : \mathbb{R}^{12} \to \mathbb{R}^{12} \ \text{ with } Cp = (0, 0, 0, 0, p_1, p_2)^T,$$

$$\overline{D} = \text{diag}(\alpha_1 + \alpha_5, \alpha_2 + \alpha_6, 1/d_3, 1/d_4, 1/d_5, 1/d_6). \tag{11.9}$$

The reduced augmented Hamiltonian has now the expansion

$$\overline{H}_{aug}(q,p) = \frac{1}{2}\langle \begin{pmatrix} q \\ p \end{pmatrix}, \begin{pmatrix} 0 & -C \\ -C^* & \overline{D} \end{pmatrix} \begin{pmatrix} q \\ p \end{pmatrix} \rangle + \mathcal{O}(|(q,p)|^3).$$

This corresponds exactly to formula (6.14), where the matrix C corresponds to C_{e_6} of Example 2 in Section 6.7 (note the different basis there).

The reduced Hamiltonian $\overline{H}$ without augmentation is obtained by adding N_0 which results in

$$\overline{H}(q,p) = p_3 + \frac{1}{2}\langle p, \overline{D}p \rangle + \mathcal{O}(|(q,p)|^3).$$

The inverse Legendre transform is possible if α_i is chosen such that $\alpha_1 + \alpha_5$, $\alpha_2 + \alpha_6 \neq 0$. The associated reduced Lagrangian $\overline{L}$ reads

$$\overline{L}(q,\dot{q}) = \frac{1}{2}\langle (\dot{q} - e_3), \overline{D}^{-1}(\dot{q} - e_3) \rangle + \mathcal{O}(|(q,\dot{q})|^3).$$

We recall that $\overline{H}$ and $\overline{L}$ can be made G–invariant, but this property is not visual when doing the expansion.

We now want to find the projected Lagrangian which is obtained by evaluating the full Lagrangian on the center manifold. We do this by expressing $L(\phi, \dot{\phi}) = W(y, I + (\nabla_y \phi, \dot{\phi}))$ in terms of (ϕ, ψ). We obtain

$$\underline{L}(\phi,\psi) = \frac{1}{2}\int_\Sigma \{\ \psi_1^2 + \psi_2^2 + \lambda_1^{-1}\psi_3^2 + 2\phi_{1,1}^2 + 2\phi_{2,2}^2$$
$$+2\lambda\lambda_2^{-1}(\phi_{1,1}+\phi_{2,2})^2 + (\phi_{1,2}+\phi_{2,1})^2\}dy + \mathcal{O}(|(\phi,\psi)|^3).$$

Inserting the center manifold $(\phi, \psi) = \sum(q_i x_i + p_i \overline{z}_i) + \mathcal{O}(|(q,p)|^2)$ yields

$$\widehat{L}_P(q,p) = \frac{1}{2}\langle p, \Gamma p \rangle + \mathcal{O}(|(q,p)|^3),$$

$$\text{with}\quad \Gamma = \text{diag}(\rho_1, \rho_2, d_3, d_4, d_5, d_6),$$
$$\rho_1 = 1/(4\lambda_1)^2 \int_\Sigma \{(\Psi_{1,1} + \lambda(y_1^2 - y_2^2))^2 + (\Psi_{1,2} + 2\lambda y_1 y_2)^2\}\, dy > 0,$$
$$\text{and}\quad \rho_2 = 1/(4\lambda_1)^2 \int_\Sigma \{(\Psi_{2,1} + 2\lambda y_1 y_2)^2 + (\Psi_{2,2} - 2\lambda(y_1^2 - y_2^2))^2\}\, dy > 0.$$

Substituting $p = \overline{D}^{-1}(\dot{q} - e_3)$ we find the projected Lagrangian

$$L_P(q,\dot{q}) = \frac{1}{2}\langle (\dot{q} - e_3), \overline{D}^{-1}\Gamma\overline{D}^{-1}(\dot{q} - e_3) \rangle + \mathcal{O}(|(q,\dot{q})|^3).$$

Thus, L_P and $\overline{L}$ coincide up to third order if and only if $\overline{D} = \Gamma$. According to (11.8) this is achieved by

$$\alpha_k = (\rho_k + \kappa_k/d_k^2)/2, \quad \alpha_{k+4} = (\rho_k - \kappa_k/d_k^2)/2, \quad k = 1,2.$$

Altogether, we have shown that natural reduction is valid in this rod problem, at least up to third order. In G–invariant form we have

$$W_{\mathrm{rod}}(R^T \dot{r}, R^T \dot{R}) = \int_\Sigma W(y, F(y, R^T \dot{r}, R^T \dot{R})) \, dy + \mathcal{O}(|(R^T \dot{r} - e_3, R^T \dot{R} - I)|^3),$$

where the function $F : \mathbf{g} \to L(I\!R^3)$ is given by the center manifold reduction. The resulting stored–energy function of the rod model is locally non–negative and convex since all diagonal elements of Γ are positive.

Finally, we want to show, that on the quadratic level there are other choices of coordinates which respect the symmetry under G and allow for natural reduction. In order to keep the symmetry we are only interested canonical coordinates (in the center space) which maintain the vectors x_i in the new symplectic basis. Hence, we have $q = \widehat{q} + M\widehat{p}$, $p = \widehat{p}$ and M has to be symmetric, where $(\widehat{q}, \widehat{p})$ are the new canonical coordinates. The new Hamiltonian $\widehat{H}$ takes the form $\widehat{H}(\widehat{q}, \widehat{p}) = \widehat{p}_3 + \frac{1}{2}\langle \widehat{p}, \widehat{D}\widehat{p}\rangle + \mathcal{O}(|(\widehat{q}, \widehat{p})|^3)$, where $\widehat{D} = \overline{D} - MC - C^*M$. As above we obtain for the reduced Lagrangian $\widehat{L}(\widehat{q}, \dot{\widehat{q}}) = \frac{1}{2}\langle(\dot{\widehat{q}} - e_3), \widehat{D}^{-1}(\dot{\widehat{q}} - e_3)\rangle$+h.o.t. and for the projected Lagrangian $L_P(\widehat{q}, \dot{\widehat{q}}) = \frac{1}{2}\langle(\dot{\widehat{q}} - e_3), \widehat{D}^{-1}\Gamma\widehat{D}^{-1}(\dot{\widehat{q}} - e_3)\rangle$+h.o.t. Imposing the natural reduction gives $\widehat{D} = \Gamma$ which was exactly the case above. Note, however, that this does not imply that M has to vanish, only $MC + C^*M = 0$ is required. Because of the special form of C (see (11.9)) there are many possible M. Thus, the canonical coordinates giving rise to natural reduction are not unique, but the associated reduced Lagrangian is always the same up to third order.

11.5 Reduction via symmetry

The Hamiltonian formulation for the *rod equations* on $T^\circ G = G \times \mathbf{g}^*$ is obtained by defining the conjugate variables

$$m = \frac{\partial}{\partial u} W_{\mathrm{rod}}(u, v), \quad n = \frac{\partial}{\partial v} W_{\mathrm{rod}}(u, v),$$

cf. (6.7). Here, n is the contact force and m the contact moment, both in coordinates with respect to the current cross–section. The Legendre transformation yields

$$u = \frac{\partial}{\partial m} \widetilde{H}(m, n), \quad v = \frac{\partial}{\partial n} \widetilde{H}(m, n).$$

Hence, we have $\mathcal{F} = Rn$ and $\mathcal{M} = Rm + r \times \mathcal{F}$.

Using the definition of the canonical symplectic structure on $T^\circ G$ given in (5.8) we find that the Hamiltonian system on the center manifold can be written as

$$\dot{r} = R\frac{\partial \widetilde{H}}{\partial n}(m, n), \quad \dot{R} = RU(m, n),$$

$$\dot{n} = n \times \frac{\partial \widetilde{H}}{\partial m}(m, n), \quad \dot{m} = m \times \frac{\partial \widetilde{H}}{\partial m}(m, n) + n \times \frac{\partial \widetilde{H}}{\partial n}(m, n),$$

where U is the skew–symmetric matric corresponding to $u = \frac{\partial \widetilde{H}}{\partial m}(m, n)$. The factorization in the sense of Section 5.2 of the twelve–dimensional systems leads to the six–dimensional system on the Lie algebra $\mathbf{g}^*$ equipped with the Lie–Poisson bracket (see Section (6.7)).

The classical reduction via symmetry described in Section 5.1 yields now the following. The momemtum map $\mathcal{J} : T^*G \to \mathbf{g}^*$ is given by $(\mathcal{F}, \mathcal{M})$, where $\mathcal{M}$ should be identified with a skew–symmetric 3×3–matrix. Factorization with respect to G can be understood in just forgetting the decoupled variables (r, R). Restricting the system to surfaces of constant values of the momemtum map $\mathcal{F} = F_0$ and $\mathcal{M} = M_0$ leads to a reduced phase space which can be given in the form

$$P_{F_0, M_0} = \{ (n, m) \in \mathbb{R}^6 \; : \; |n| = |F_0|, \; n \cdot m = F_0 \cdot M_0 \} \quad \text{if} \quad F_0 \neq 0,$$
$$P_{0, M_0} = \{ (0, m) \in \mathbb{R}^6 \; : \; |m| = |M_0| \} \quad\quad\quad\quad \text{if} \quad F_0 = 0.$$

In particular, the reduced phase space is two–dimensional if $F_0 = 0$ and four–dimensional else. For each of these reduced phase spaces canonical coordinates are given in [HM88].

In the case of $|F_0|$ sufficienctly small the existence of transverse homoclinic points, and hence of chaotic dynamics, is shown to exist on P_{F_0, M_0}.

11.6   Cross–sectional symmetries and reversibility

Finally we want to show how cross–sectional symmetries penetrate from the beam equations to the rod equations. Therefore we refer back to the example given in Section 5.2, page 44, which reduces to the beam problem when letting $n = 3$ and $m = 3$.

The compact Lie group S is a subgroup of $O(2) = \{ \tau \in \mathbb{R}^{2 \times 2} \; : \; \tau^* \tau = I \}$ acting on $y \in \Sigma \subset \mathbb{R}^2$ such that $\tau \Sigma = \Sigma$. Moreover, S acts on $(\varphi, \dot{\varphi})$ through Φ_τ given by

$$\Phi_\tau (\varphi, \dot{\varphi})(y) = \left(\begin{pmatrix} \tau & 0 \\ 0 & 1 \end{pmatrix} \varphi(\tau^* y), \begin{pmatrix} \tau & 0 \\ 0 & 1 \end{pmatrix} \dot{\varphi}(\tau^* y) \right).$$

Hence, if Σ is a disk we have $S = O(2)$ and otherwise S is a finite group, a subgroup of a dihedral group D_n. The case of two orthogonal symmetry axis gives $S = \mathbb{Z}_2 \times \mathbb{Z}_2$ and was already treated in [Mi88c].

Additionally the system can be reversible which corresponds to the reflection $t \to -t$ and the mapping

$$\mathcal{R}(\varphi, \dot{\varphi}) = (\Gamma \varphi, -\Gamma \varphi) \quad \text{with } \Gamma \varphi = (\varphi_1, \varphi_2, -\varphi_3);$$

see Section 5.5. Note here that $\Phi_\tau(\mathcal{R}(\varphi, \dot{\varphi})) = \mathcal{R}(\Phi_\tau(\varphi, \dot{\varphi}))$ for all $\tau \in S$.

However, we also have to check whether the Lagrangian $L(\varphi, \dot{\varphi}) = \int_\Sigma W(\nabla_y \varphi, \dot{\varphi}) \, dy$ is invariant under S and $\mathcal{R}$. Therefore let $(\tilde{\varphi}, \tilde{\dot{\varphi}}) = \Phi_\tau(\varphi, \dot{\varphi})$, then the strain tensor transforms $E = \mathcal{E}(\nabla\varphi) = \frac{1}{2}(\nabla\varphi^* \nabla\varphi - I)$ transforms into

$$\widetilde{E} = \mathcal{E}(\nabla_y\widetilde{\varphi}, \dot{\widetilde{\varphi}}) = \begin{pmatrix} \tau & 0 \\ 0 & 1 \end{pmatrix} \mathcal{E}(\nabla_y\varphi, \dot{\varphi}) \begin{pmatrix} \tau^* & 0 \\ 0 & 1 \end{pmatrix}.$$

Similarly, for $(\widetilde{\varphi}, \dot{\widetilde{\varphi}}) = \mathcal{R}(\varphi, \dot{\varphi})$ we find $\widehat{E} = \Gamma E \Gamma$. For isotropic material we have $\widetilde{W}(QEQ^*) = \widetilde{W}(E)$ for all $Q \in O(3)$, see [Ba77]. Here we need this property just for the subgroup $S \odot \mathbb{Z}_2 = \{ \begin{pmatrix} \tau & 0 \\ 0 & \pm 1 \end{pmatrix} \in O(3) : \tau \in S \}$. For inhomogeneous materials, depending on the cross–sectional variable y, we have to impose

$$\widetilde{W}(\tau y, E) = \widetilde{W}(y, \begin{pmatrix} \tau & 0 \\ 0 & \pm 1 \end{pmatrix} E \begin{pmatrix} \tau^* & 0 \\ 0 & \pm 1 \end{pmatrix})$$

for all τ, y, and E. For a general discussion of material and domain symmetries in elasticity problems we refer to [He88] and the references therein.

Since the Theorems of Chapter 5 are applicable to the joint action of $\widetilde{G} = G \odot S$ and the reversibility operator $\mathcal{R}$ we know that the reduced Lagrangian $W_{\text{rod}} = W_{\text{rod}}(u, v)$ is again invariant under the reduced actions of S and $\mathcal{R}$. The reduced actions are linear representations given by the actions on the tangent space of the center manifold. According to Section 11.3 the tangent space of the factorized center manifold, given over **g** only, is spanned by the six vectors $\widehat{z}_1, \ldots, \widehat{z}_6$. Using the coordinates $(u, v) = (\dot{q}_4, \dot{q}_5, \dot{q}_6, \dot{q}_1, \dot{q}_2, \dot{q}_3) + \mathcal{O}(|\dot{q}|^2)$ in this subspace and working out the actions on the vectors $\widehat{z}_j$ we easily find the linear representations

$$\tau : (u, v) \rightarrow (\begin{pmatrix} (\det \tau)\tau & 0 \\ 0 & 1 \end{pmatrix} u, \begin{pmatrix} \tau & 0 \\ 0 & 1 \end{pmatrix} v),$$

$$\mathcal{R} : (u, v) \rightarrow (\Gamma u, -\Gamma v).$$

Hence, the invariance condition for W_{rod} is given by

$$W_{\text{rod}}(u, v) = W_{\text{rod}}(\begin{pmatrix} (\det \tau)\tau & 0 \\ 0 & \pm 1 \end{pmatrix} u, \begin{pmatrix} \pm\tau & 0 \\ 0 & 1 \end{pmatrix} v)$$

for all $\tau \in S$ and "$-$" for the reversible case only.

Bibliography

[AK89] C.J. AMICK AND K. KIRCHGÄSSNER. A theory of solitary water–waves in the presence of surface tension. *Arch. Rational Mech. Analysis*, **105**, 1–49, 1989.

[AM78] R. ABRAHAM AND J.E. MARSDEN. *Foundations of Mechanics*. Benjamin / Cummings, Reading, MA, 1978.

[An72] S. ANTMAN. The theory of rods. *Handbuch der Physik Vol. VIa*, 1972. Springer-Verlag.

[Ar78] V.I. ARNOLD. *Mathematical Methods in Classical Mechanics*. Springer–Verlag, New York, 1978. Grad. Texts Math. Vol. 60.

[Au85] B. AULBACH. A classical approach to the analyticity problem of center manifolds. *J. Applied Math. Physics (ZAMP)*, **36**, 1–23, 1985.

[Ba77] J.M. BALL. Constitutive inequalities and existence theorems in nonlinear elastostatics. In R.J. Knops, editor, *Nonlinear Analysis and Mechanics: Heriot-Watt Symposium Vol. I*, pages 187–241, Pitman, London, 1977.

[BB87] T.B. BENJAMIN AND S. BOWMAN. Discontinuous solutions of one–dimensional Hamiltonian systems. *Proc. Roy. Soc. London, Ser.A*, **413**, 263–295, 1987.

[BC74] N. BURGOYNE AND R. CUSHMAN. Normal forms for real linear Hamiltonian systems with purely imaginary eigenvalues. *Celestial Mech.*, **8**, 435–443, 1974.

[BC77] N. BURGOYNE AND R. CUSHMAN. Normal forms for real linear Hamiltonian systems. In C. Martin and R. Hermann, editors, *The 1976 Ames Research Center (NASA) Conference on Geometric Control Theory*, pages 483–529, Math Sci Press, Brookline (Mass), 1977.

[BM90] C. BAESENS AND R.S. MacKAY. Uniformly travelling water waves from a dynamical systems viewpoint: some insight into bifurcations from the Stokes family. Manuscript, Nonlinear Systems Laboratory, Univ. Warwick, 1990.

[BO82] T.B. BENJAMIN AND P.J. OLVER. Hamiltonian structure, symmetries and conservation laws for water waves. *J. Fluid Mechanics*, **125**, 137–185, 1982.

[Br88] A.D. BRYUNO. The normal form of a Hamiltonian system. *Russian Math. Surveys*, **43**, 25–66, 1988.

[Br89] A.D. BRYUNO. *Local Methods in Nonlinear Differential Equations*. Springer–Verlag, 1989.

[Ca81] J. CARR. *Applications of Centre Manifold Theory*. Springer–Verlag, 1981. Appl. Math. Sci. Vol. 35.

[CE90] P. COLLET AND J.–P. ECKMANN. The time dependent amplitude equation for the Swift–Hohenberg problem. *Comm. Math. Physics*, **132**, 139–153, 1990.

[CH82] S.-N. CHOW AND J.K. HALE. *Methods of Bifurcation Theory*. Springer–Verlag, 1982.

[CM74] P.R. CHERNOFF AND J.E. MARSDEN. *Properties of Infinite Dimensional Hamiltonian Systems*. Springer-Verlag, Berlin New-York Heidelberg, 1974. Lecture Notes Math. Vol. 425.

[Da89] B. DACOROGNA. *Direct Methods in the Calculus of Variations*. Springer–Verlag, 1989. Applied Math. Sciences Vol.78.

[DES71] R.C. DIPRIMA, W. ECKHAUS, AND L.A. SEGEL. Non–linear wave–number interaction in near–critical two–dimensional flows. *J. Fluid Mechanics*, **49**, 705–744, 1971.

[dS55] A.J.CB. DE SAINT–VENANT. Mémoire sur la torsion des prismes. *Mémoires preésntés pars divers savant a l'Académie de l'Institut Impérial de France Ser.2*, **14**, 233–560, 1855.

[Er77] J. ERICKSEN. On the formulation of Saint-Venant's problem. In R.J. Knops, editor, *Nonlinear Analysis and Mechanics: Heriot-Watt Symposium Vol. I*, pages 158–186, Pitman, London, 1977.

[ETB*87] C. ELPHICK, E. TIRAPEGUI, M.E. BRACHET, P. COULLET, AND G. IOOSS. A simple global characterization of normal forms of singular vector fields. *Physica D*, **29**, 95–127, 1987.

[Fi84] G. FISCHER. Zentrumsmannigfaltigkeiten bei elliptischen Differentialgleichungen. *Math. Nachr.*, **115**, 137–157, 1984.

[GH83] J. GUCKENHEIMER AND P. HOLMES. *Nonlinear Oscillations, Dynamical Systems, and Bifurcations of Vector Fields*. Springer–Verlag, 1983. Appl. Math. Sciences Vol. 42.

[Gr86] M. GRMELA. Bracket formulation of diffusion-convection equations. *Physica D*, **21**, 179–212, 1986.

[GS84] V. GUILLEMIN AND S. STERNBERG. *Symplectic Techniques in Physics.* Cambridge University Press, 1984.

[GSS88] M. GOLUBITSKY, I. STEWART, AND D.G. SCHAEFFER. *Singularities and Groups in Bifurcation Theory, Vol. II.* Springer–Verlag, 1988. Appl. Math. Sci. Vol. 69.

[He81] D. HENRY. *Geometric Theory of Semilinear Parabolic Equations.* Springer–Verlag, 1981. Lect. Notes Math. Vol. 840.

[He88] T.J. HEALEY. Global bifurcation and continuation in the presence of symmetry with an application to solid mechanics. *SIAM J. Math. Analysis,* **19**, 824–840, 1988.

[HKM77] T.J.R. HUGHES, T. KATO, AND J.E. MARSDEN. Well-posed quasi-linear hyperbolic systems with applications to nonlinear elastodynamics and general relativity. *Arch. Rational Mech. Analysis,* **63**, 273–294, 1977.

[HM83] T.J.R. HUGHES AND J.E. MARSDEN. *Mathematical Foundations of Elasticity.* Prentice Hall, Englewoods Cliffs NJ, 1983.

[HM88] P.J. HOLMES AND A. MIELKE. Spatially complex equilibria of buckled rods. *Arch. Rational Mech. Analysis,* **101**, 319–348, 1988.

[Ho86] P. HOLMES. Chaotic motions in a weakly nonlinear model for surface waves. *J. Fluid Mech.,* **162**, 365–388, 1986.

[HS85] J.K. HALE AND J. SCHEURLE. Smoothness of bounded solutions of nonlinear evolution equations. *J. Diff. Eqns.,* **56**, 142–163, 1985.

[IA91] G. IOOSS AND M. ADELMEYER. Topics in Bifurcation Theory and Applications. Lecture Notes of a course at Universität Stuttgart, 1991.

[IK91] G. IOOSS AND K. KIRCHGÄSSNER. Water waves for small surface tension – an approach via normal form. *Proc. Roy. Soc. Edinburgh Ser.A,* 1991. To appear.

[IM91] G. IOOSS AND A. MIELKE. Bifurcating time–periodic solutions of Navier–Stokes equations in infinite cylinders. *J. Nonlinear Sciences,* **1**, 107–146, 1991.

[IV91] G. IOOSS AND A. VANDERBAUWHEDE. Center manifold theory in infinite dimensions. *Dynamics Reported,* 1991. To appear.

[Ke67] A. KELLEY. The stable, center–stable, center, center–unstable, unstable manifolds. *J. Diff. Eqns.,* **3**, 546–570, 1967.

[Ki59] G. KIRCHHOFF. Über das Gleichgewicht und die Bewegungen eines unendlich dünnen Stabes. *J. für Mathematik (Crelle),* **56**, 285–313, 1859.

[Ki82] K. KIRCHGÄSSNER. Wave solutions of reversible systems and applications. *J. Diff. Eqns.*, **45**, 113–127, 1982.

[Ki88] H. KIELHÖFER. A bifurcation theorem for potential operators. *J. Funct. Anal.*, **77**, 1–8, 1988.

[Ki91] P. KIRRMANN. Zentrumsmannigfaltigkeiten für voll nichtlineare elliptische Gleichungen. Thesis Universität Stuttgart, 1991.

[Kl82] W. KLINGENBERG. *Riemannian Geometry.* de Gruyter, Berlin-Heidelberg, 1982.

[KMS88] P. KRISHNAPRASAD, J.E. MARSDEN, AND J. SIMO. The Hamiltonian Structure of Nonlinear Elasticity: The convective representation of solids, rods, and plates. *Arch. Rational Mech. Analysis*, **104**, 125–184, 1988.

[Kr90] M. KRUPA. Bifurcations of relative equilibria. *SIAM J. Math. Anal.*, **21**, 1453–1486, 1990.

[KS77] J.K. KNOWLES AND E. STERNBERG. On the failure of ellipticity of the equations for finite elastostatic plane strain. *Arch. Rational Mech. Analysis*, **63**, 321–336, 1977.

[La62] S. LANG. *Introduction to Differentiable Manifolds.* John Wiley & Sons, 1962.

[Ma81] J.E. MARSDEN. *Lectures on Geometric Methods in Mathematical Physics.* SIAM, 1981. Reg. Conf. Series in Applied Math. Vol.37.

[MHO91] A. MIELKE, P.J. HOLMES, AND O. O'REILLY. Cascades of homoclinic orbits to, and chaos near, a Hamiltonian saddle–center. *J. Dynamics Diff. Eqns.*, 1991. To appear.

[Mi86a] A. MIELKE. A reduction principle for nonautonomous systems in infinite-dimensional spaces. *J. Diff. Eqns.*, **65**, 68–88, 1986.

[Mi86b] A. MIELKE. Steady flows of inviscid fluids under localized perturbations. *J. Diff. Eqns.*, **65**, 89–116, 1986.

[Mi88a] A. MIELKE. On Saint-Venant's problem for an elastic strip. *Proc. Roy. Soc. Edinburgh*, **110A**, 161–181, 1988.

[Mi88b] A. MIELKE. Reduction of quasilinear elliptic equations in cylindrical domains with applications. *Math. Meth. Appl. Sci.*, **10**, 51–66, 1988.

[Mi88c] A. MIELKE. Saint-Venant's problem and semi-inverse solutions in nonlinear elasticity. *Arch. Rational Mech. Analysis*, **102**, 205–229, 1988.

[Mi89] A. MIELKE. On nonlinear problems of mixed type: a qualitative theory using infinite-dimensional center manifolds. *J. Dynamics Diff. Eqns.*, 1989. Submitted.

[Mi90] A. MIELKE. Normal hyperbolicity of center manifolds and Saint-Venant's principle. *Arch. Rational Mech. Analysis*, **110**, 353–372, 1990.

[Mi91a] A. MIELKE. Locally invariant manifolds for quasilinear parabolic equations. *Rocky Mountain J. Math.*, **21**, 1991. No. 2.

[Mi91b] A. MIELKE. Reduction of PDEs on domains with several unbounded directions: a step towards modulation equations. *J. Applied Math. Physics (ZAMP)*, 1991. Submitted.

[Mo58] J. MOSER. On the generalization of a theorem of A. Liapunoff. *Comm. Pure Appl. Math.*, **11**, 257–271, 1958.

[Mo77] J. MOSER. Proof of a generalized form of a fixed point theorem due to G.D. Birkhoff. In J. Palis and M. do Carmo, editors, *Geometry and Topology*, pages 464–494, Springer–Verlag, 1977. Lect. Notes Math. Vol. **597**.

[MR86] J.E. MARSDEN AND T. RATIU. Reduction of Poisson manifolds. *Letters Math. Physics*, **11**, 161–169, 1986.

[MRS90] J. MONTALDI, M. ROBERTS, AND I. STEWART. Existence of nonlinear normal modes of symmetric Hamiltonian systems. *Nonlinearity*, **3**, 695–730, 1990.

[MSLP89] J.E. MARSDEN, J.C. SIMO, D. LEWIS, AND T.A. POSBERGH. Block diagonalization and the energy–momentum method. *Contemporary Math. (AMS)*, **97**, 297–313, 1989.

[NW69] A. NEWELL AND J. WHITEHEAD. Finite bandwidth, finite amplitude convection. *J. Fluid Mechanics*, **38**, 279–303, 1969.

[Ow87] N. OWEN. Existence and Stability of Necking Deformations for Nonlinearly Elastic Rods. *Arch. Rational Mech. Analysis*, **98**, 357–384, 1987.

[OY83] O.A. OLEINIK AND G.A. YOSIFIAN. On the asymptotic behavior at infinity of solutions in linear elasticity. *Arch. Rational Mech. Analysis*, **78**, 29–53, 1983.

[Pl64] V.A. PLISS. A reduction principle in the theory of stability of motion. *Izv. Akad. Nauk. SSSR Ser. Mat.*, **28**, 1297–1324, 1964.

[Po80] J. PÖSCHEL. Über invariante Tori in differenzierbaren Hamiltonschen Systemen. *Bonner Math. Schriften*, **120**, 1–103, 1980.

[RT71] D. RUELLE AND F. TAKENS. On the nature of turbulence. *Commun. Math. Phys.*, **20**, 167–192, 1971. See also **23**, 343–344, 1971.

[Ru73] D. RUELLE. Bifurcations in the presence of a symmetry group. *Arch. Rational Mech. Analysis*, **51**, 136–152, 1973.

[Sc91] B. SCARPELLINI. Center manifolds of infinite dimensions. *J. Applied Math. Physics (ZAMP)*, 1991. To appear.

[Si90] C. SIMO. Study of space craft orbits via central manifold. Lecture given at the *Workshop on Dynamical Systems*, CRM, Barcelona, October 1–5, 1990.

[SS83] H.C. SIMPSON AND S.J. SPECTOR. On copositive matrices and strong ellipticity for isotropic elastic materials. *Arch. Rational Mech. Analysis*, **84**, 55–68, 1983.

[vH90] A. VAN HARTEN. On the validity of Ginzburg–Landau's equation. University Utrecht, Preprint 619, 1990.

[We71] A. WEINSTEIN. Symplectic manifolds and their Lagrangian submanifolds. *Advances in Math.*, **6**, 329–346, 1971.

[Wi36] J. WILLIAMSON. On the algebraic problem concerning the normal forms of linear dynamical systems. *Amer. J. Math.*, **58**, 141–163, 1936.

Index